CONTENTS

AUTHOR'S PREFACE

This book is primarily intended for use by students who are engaged in courses leading to the awards of Ordinary and Advanced Level Certificates, Engineering Degrees, Higher Technician Certificates and Diplomas, Technician Certificates and Diplomas, and Membership of the Professional Engineering Institutions. The questions have therefore been selected principally from past papers issued by these Examining Authorities, although the time-allocations and solutions are the author's own.

It will be appreciated that a variation in standard exists between different examinations and by careful selection of questions and the inclusion of private work provision has been made to ensure that the questions have been graded progressively from a standard slightly below Ordinary Level G.C.E. to that of Engineering Degree Examinations. A serious attempt has been made to cover the complete range of questions that occur in engineering drawing examinations between these two standards.

A principal feature of the book is the provision of model solutions whereby the student can correct his own work with the minimum of supervision. Therefore notes and, in certain examples, pictorial projections have been added as an aid for comprehension of the given solutions. In the author's opinion worked examples are a valuable method of instruction not normally provided in books on engineering drawing.

Generally examination questions tend to be rather academic in their presentation and not usually in accordance with normal Drawing Office Practice. The inclusion of some examples of industrial type drawings should help the student to appreciate this difference. All the drawings are to BS 308 (1972) and relevant BS Data Sheets have been included.

I gratefully acknowledge permission given by the following bodies to reproduce questions from their examination papers in Engineering Drawing: The Senate of the University of London: The Institution of Mechanical Engineers; The Part I Committee of the Joint Engineering Institutions; The City and Guilds of London Institute; The West African Examinations Council; U.S.T. Kumasi; Enfield College; The British Standards Institution. I would also like to record my sincere thanks for the help that I received from my colleagues Mr. D. J. Clarke, B.Sc. (Hons.) Eng., C.Eng., M.I.Mech.E., M.R.I.N.A., A.M.I.E.D., and Mr. T. Keating, C.Eng., M.I.Mech.E.

K.R.H.

1 METRICATION

The Confederation of British Industries decided early in 1965 to recommend to the government that British Industry adopt the metric system. The government was impressed with the case put to it by the representatives of industry and gave its full support, indicating its hope that by 1975 the greater part of the country's industry would have effected the change.

The metric system chosen was the International System of Units (officially designated SI Units), and the British Standards Institution is engaged on a priority task of preparing some 1400 standards.

It was envisaged that design and drawing offices of British Industry would introduce metric sizes during 1968 and 1969 and would have completed this transition from imperial sizes by 1970 (see *Change to the metric system* published by EEUA). The Royal Society and CEI have also recommended that SI units be increasingly used in university and college teaching and in particular that numerical data appearing in examinations set to those students now entering courses of higher education be quoted in SI units.

Use of SI Units in Engineering Drawing

As far as engineering drawing is concerned, the use of SI units means that all dimensions will be stated in millimetres, or in metres on drawings of very large equipment.

When new designs are being developed it is important that the designer thinks in terms of metric sizes rather than inch sizes and their subsequent conversion. New designs will have to be prepared on a metric basis, using customary metric sizes and metric components, and taking account of internationally agreed metric standards and the practice of the principal metric countries. For existing designs which are to be converted to metric sizes BSI have made available a standard which provides a procedure for conversion which gives the essential accuracy required for precise dimensional interchangeability.

The drawings in this book have been selected from past papers set by various examining authorities and were dimensioned using inch units. These have been converted to the nearest millimetre with a certain amount of 'rounding off', and all screw threads have been changed from the many imperial types, i.e. BSW, BS Fine, BA, etc., to the metric thread sizes outlined in BS 3643, Parts 1, 2 and 3. This is the British Standard for ISO (International Standards Organization) metric screw threads and recommends that the coarse pitch series be used for the vast majority of general purpose applications. Although the ISO unified thread will still be acceptable and widely used, a complete change to ISO metric is recommended. An ISO metric thread is denoted in the manner shown:

M12 × 1·5 – 6H

where M .... indicates that the thread is of ISO metric form.
12 .... is the thread diameter in millimetres.
1·5 .... is the thread pitch in millimetres.
6H .... indicates the thread tolerance symbols and shows that it is an internal thread. (See table below.)

Alternatively M8 × 1·0 – 6g is an external thread.

Class of fit	Tolerance class	
	Internal	External
Close	5H	4h
Medium	6H	6g
Free	7H	8g

As previously mentioned, it has been recommended that the coarse pitch series of screw threads be used for the majority of general engineering applications. These coarse series threads are designated by the symbol M and the nominal thread diameter in millimetres, e.g. M30, together with the tolerance class in the normal way but it is not necessary to include the thread pitch.

Thus M30 – 6g indicates a Coarse Series thread
and M30 × 2 – 6g indicates a Fine Series thread.

BIBLIOGRAPHY

BS 308 Engineering Drawing Practice.
Part 1, 1972: General Principles.
Part 2, 1972: Dimensioning and Tolerancing of Size.
Part 3, 1972: Geometrical Tolerancing.
BS 350, Part 1, 1959; Part 2, 1962: Conversion Factors and Tables.
BS 1957, 1953: Presentation of Numerical Values.
BS 1991, 1961–7: Letter Symbols, Signs and Abbreviations, Parts 1 to 6.
BS 2045, 1965: Preferred Numbers.
BS 2856, 1957: Precise Conversion of Inch and Metric Sizes on Engineering Drawings.
BS 3580, 1964: Guide to Design Considerations on the Strength of Screw Threads.
BS 3643, 1967: ISO Metric Screw Threads, Parts 1, 2 and 3.
BS 3692, 1967: ISO Metric Hexagon Bolts, Screws and Nuts.
BS 3763, 1964: The International System (SI) Units.
BS 4168, 1967: Hexagon Socket Screws and Wrench Keys – Metric.
BS 4183, 1967: Metric Machine Screws and Machine Screw Nuts – Metric.
BS 4186, 1967: Recommendations for Clearance Holes for Metric Bolts and Screws – Metric.

BS Handbook No. 18, 1966: Metric Standards for Engineering.
PD 5686, 1972: The Use of SI Units.
Change to the Metric System in the United Kingdom. Report by the Standing Joint Committee on Metrication, HMSO.
Changing to the Metric System. Conversion Factors, Symbols and Definitions, HMSO.
Going Metric. Poster and leaflet, Ministry of Technology.
The Adoption of the Metric System in Engineering. Basic programme and guide, BSI.
Report of the Metric Standards Working Party, Forming a Basis for the Change over from Imperial Units to Metric Units. Published by the Plessey Company Limited (inquiries to the Department of Trade and Industry).

2 INTRODUCTION TO ENGINEERING DRAWING

Drawing is the principal means of communication in engineering. It is the method used to impart ideas, convey information and specify shape and is often said to be the language of the engineer. It is an international language and bound, like any other language, by rules and conventions. These may vary slightly in detail from country to country but the underlying basic principles are common and standard.

In the United Kingdom and many other countries, British Standard 308 is the accepted authority for the rules and conventions governing engineering drawing practice. It is therefore important that the engineering student should be familiar with this publication as he will be required to exhibit this knowledge in any examination in engineering drawing.

Draughtsmanship

In general, good draughtsmanship is an art obtained by experience and improved by constant practice. Correct choice and use of instruments, layout and clarity of views, neatness and legibility of printing, etc., contribute to give a drawing character and the professional appearance associated with good draughtsmanship.

However the basic fundamental of good draughtsmanship is linework, and in student drawings poor linework is a very common feature.

BS 308: Part 1: 1972 recommends the use of different types of line as illustrated in Figure 1A. It can be seen that each type has a clearly defined application and it is strongly recommended that the student should learn and apply these recommendations.

Type A

This line represents the visible outlines and thus defines the general overall shape. It should be bold, definite and uniform in thickness and density. Homogeneity of the line is important, but arcs and circles, being more difficult to draw, tend to be thinner and less dense than straight lines. Compasses should therefore contain a softer grade of lead than that used in the pencil.

Line control is easier using a pencil and straight edge than with compasses and it is helpful to draw all arcs and circles before blending tangent lines, starting from the point of tangency.

Type B

This line is used for dimension, projection, section and leader lines, and its use is discussed under the appropriate section at a later stage.

Type C

This continuous wavy thin line is used principally to limit partial views.

Type D

This broken line represents the outline of hidden detail and is one third to one half the thickness of Type A lines.

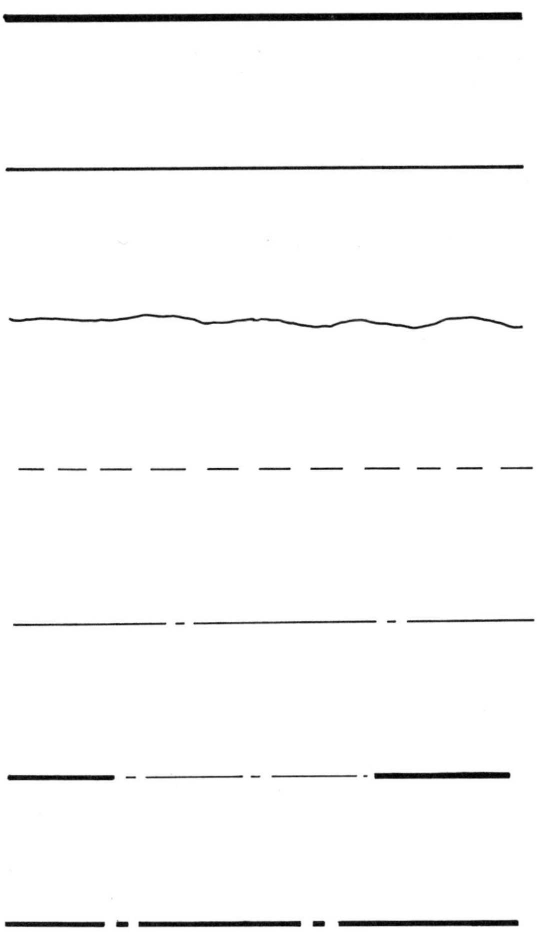

TYPE OF LINE		EXAMPLE	APPLICATION
continuous (thick)	A	━━━━━━━━━━	visible outlines and edges
continuous (thin)	B	──────────	fictitious outlines and edges dimension and leader lines hatching outlines of adjacent parts outlines of revolved sections
continuous irregular (thin)	C	～～～～～	limits of partial views or sections when the line is not an axis
short dashes (thin)	D	— — — — —	hidden outlines and edges
chain (thin)	E	——— - ——— - ———	centre lines extreme positions of moveable parts
chain (thick at ends and at changes of direction, thin elsewhere)	F	━━ - ——— - ——— - ━━	cutting planes
chain (thick)	G	━━ ▪ ━━━ ▪ ━━━	indication of surfaces which have to meet special requirements

LINES SHOULD BE SHARP AND DENSE TO OBTAIN GOOD REPRODUCTION.

LINES SPECIFIED AS THICK SHOULD BE FROM TWO TO THREE TIMES THE THICKNESS OF LINES SPECIFIED AS THIN.

CENTRE LINES SHOULD PROJECT FOR A SHORT DISTANCE BEYOND THE OUTLINE TO WHICH THEY REFER, BUT WHERE NECESSARY TO AID DIMENSIONING OR TO CORRELATE VIEWS, THEY MAY BE EXTENDED.

Figure 1A

Superimposition of hidden detail on a drawing often leads to vagueness concerning shape description. The use of this type of line involves techniques shown in Figure 1B which will clarify drawings involving complicated hidden detail.

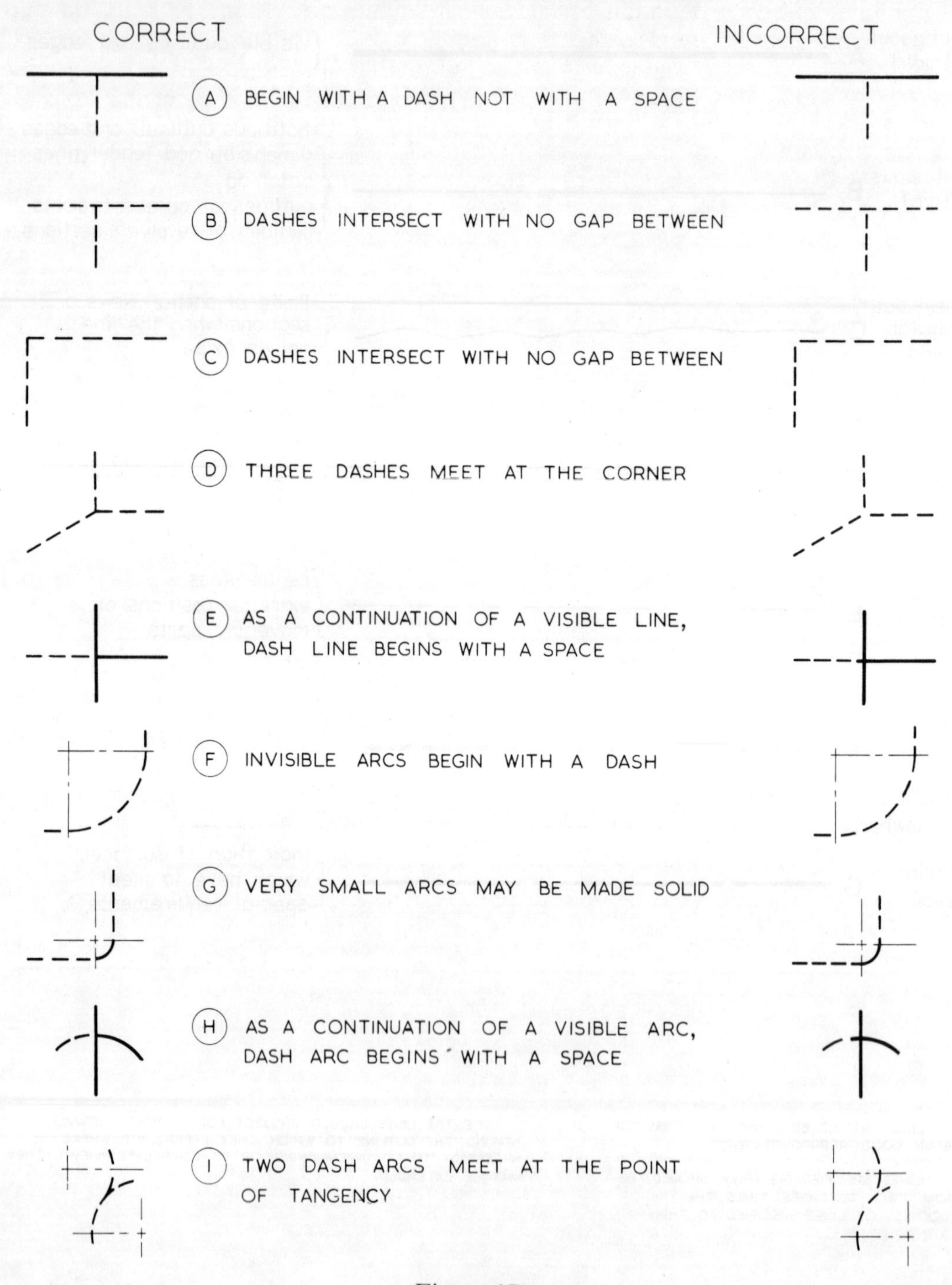

Figure 1B

Type E

This line principally represents the centre line of holes and other common features. It forms datum lines for dimensioning and axes of symmetry. Students are advised therefore to cultivate the habit of drawing these lines as in students' work they are generally most conspicuous by their absence.

Type A, B, D and E lines are in most common use and feature in all engineering drawing examinations. Figure 1C shows the typical use of all the recommended types of line; differentiation between them can only improve a student's standard of draughtsmanship, and of course, his allocation of marks for this aspect of engineering drawing examinations.

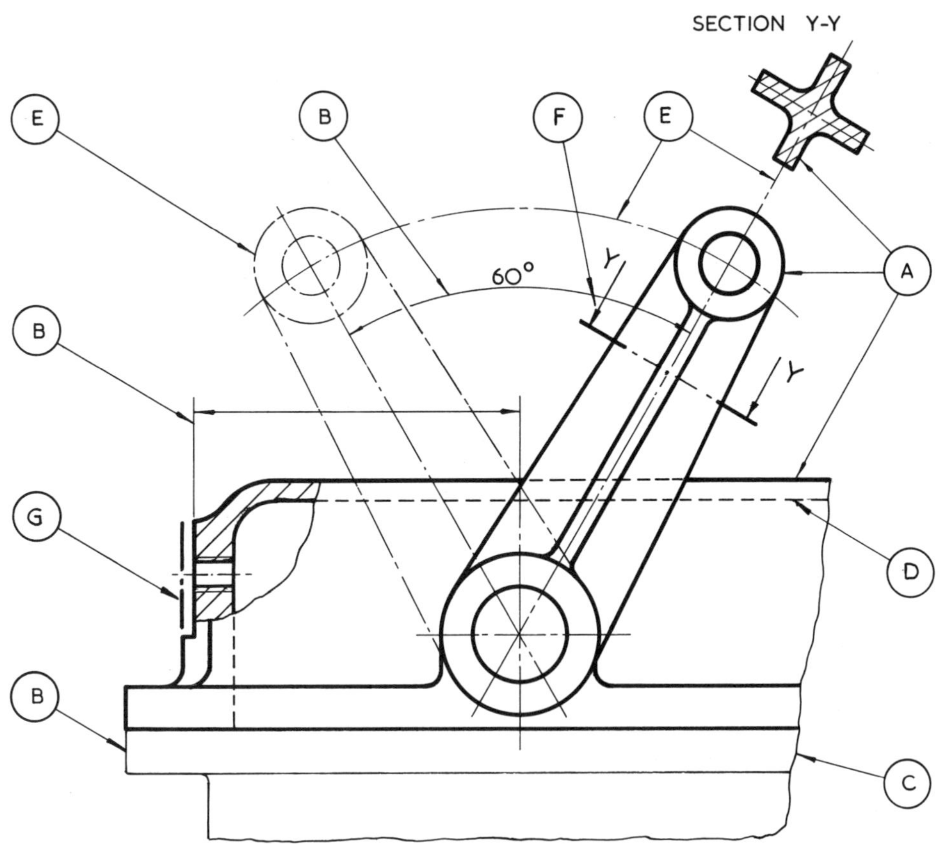

Figure 1C

3 ORTHOGRAPHIC PROJECTION

This is the method used to specify the shape description of a three dimensional object on drawing paper which is of course a two dimensional plane surface.

Projection is the representation on a plane surface of the image of an object as it is observed by a viewer. The plane on which the image is represented is known as the Plane of Projection. When the object is viewed orthogonally, i.e., at right angles to the plane of projection, then the representation of the image is said to be in orthographic projection.

In order to specify fully the shape of an object in orthographic projection at least two planes of projection are required. These principal planes of projection or reference are known as the Horizontal Plane (HP) and the Vertical Plane (VP) and are mutually perpendicular (Figure 2A). The intersection of these planes form quadrants or angles

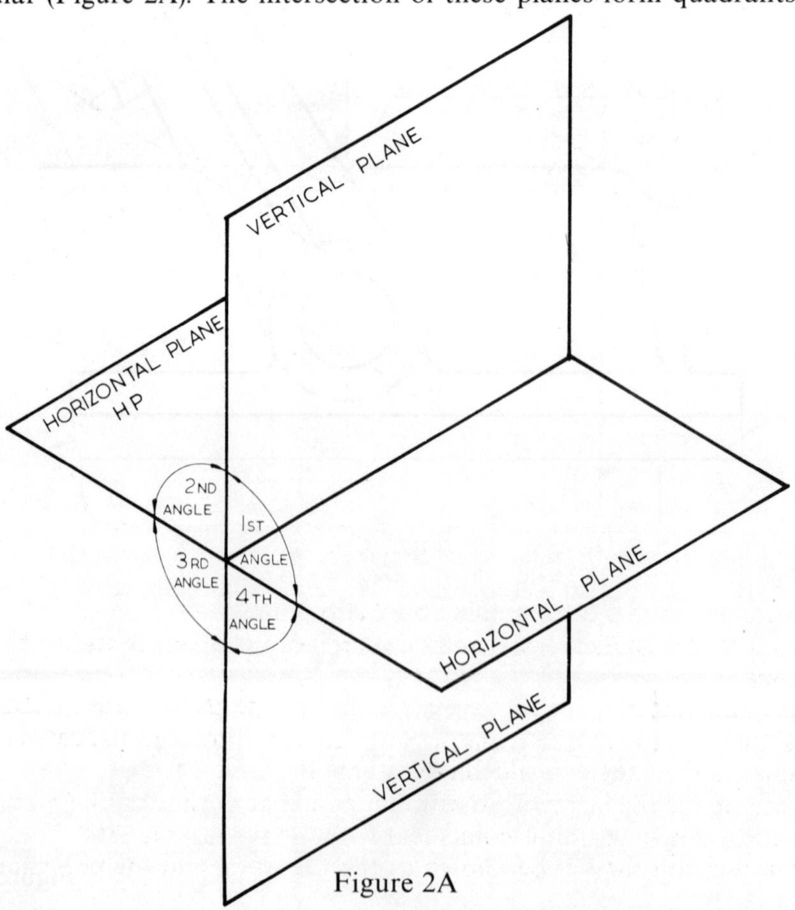

Figure 2A

of 90 degrees known as the 1st, 2nd, 3rd and 4th angles. Consider an object situated in space relative to the principal planes of reference as illustrated pictorially in Figure 2B.

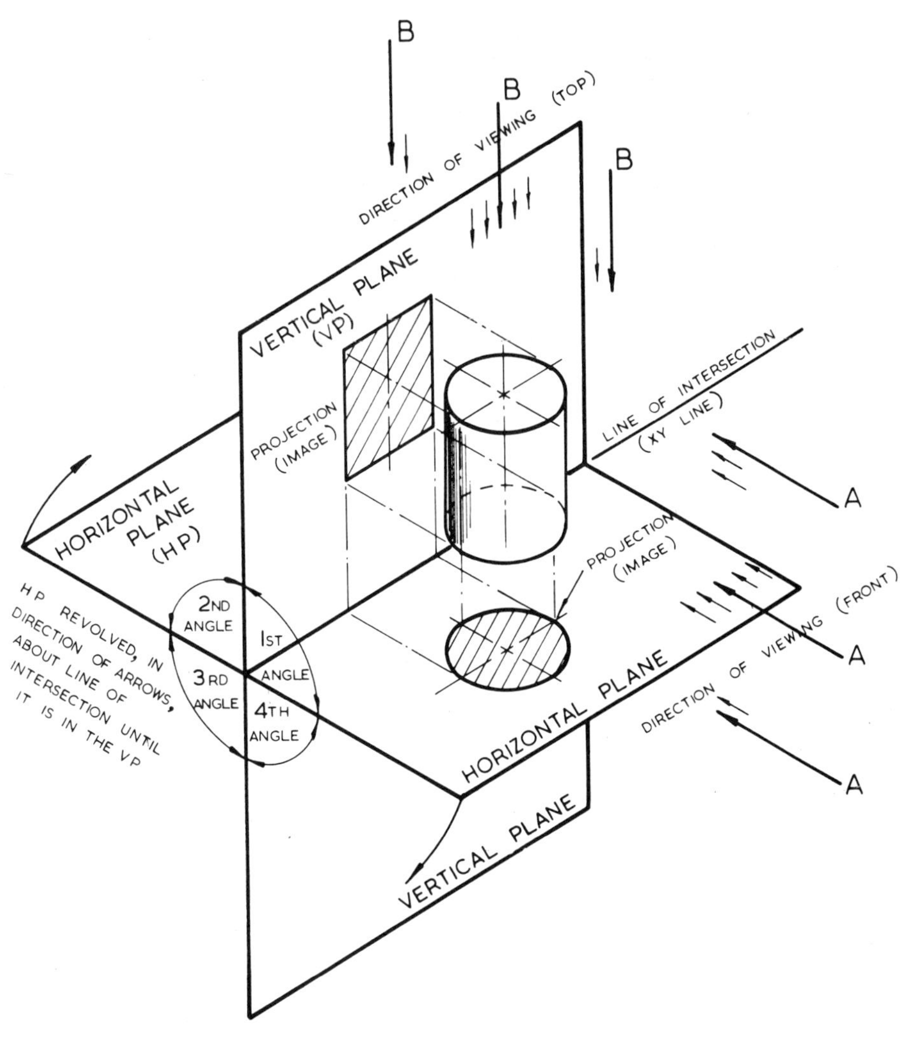

Figure 2B

Convention governing orthographic projection rules that:

1 The angles be designated as shown and therefore the object is said to be situated in the first angle.
2 The lines of sight are parallel, at right angles to the planes of projection and in general, the object is viewed from the top and front positions indicated regardless of the quadrant or angle in which it is situated.
3 The horizontal plane be revolved or rabatted about the line of intersection XY in the direction shown until it is coincident with the vertical plane.
4 The projection on the VP be known as the Elevation and the projection on the HP be known as the Plan.

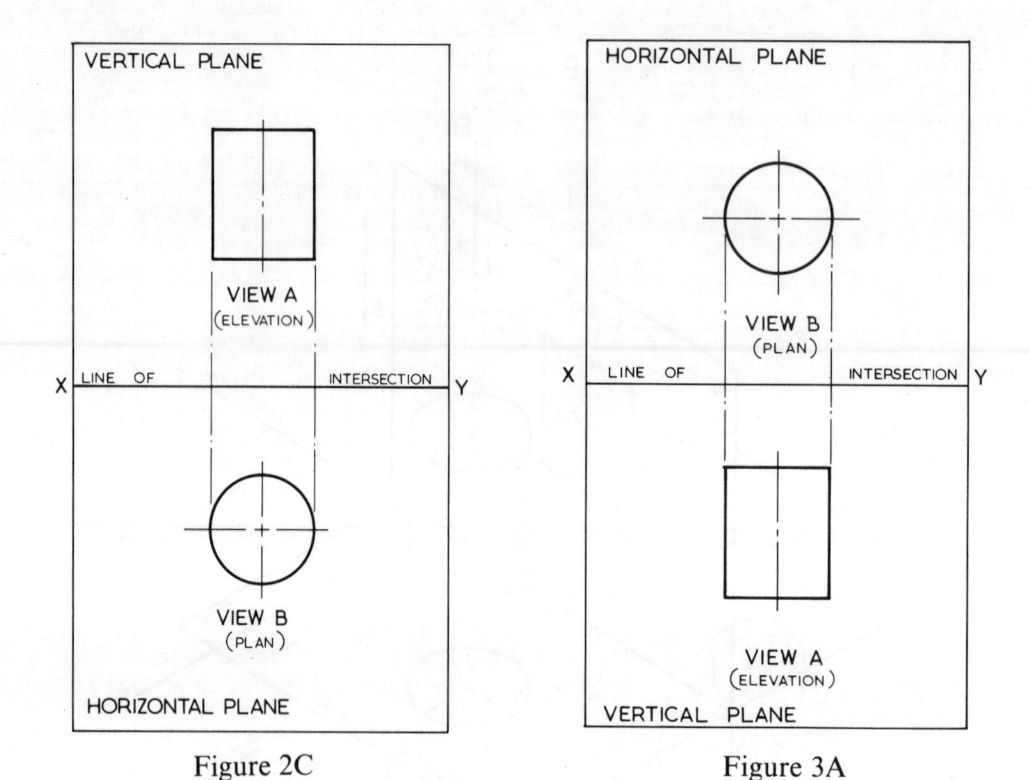

Figure 2C

Figure 3A

Therefore in accordance with this accepted convention Figure 2C shows the Plan and Elevation of the object in First Angle Orthographic Projection.

Consider the object situated in the Third Angle or quadrant (Figure 3B) and apply the same rules. It can be seen from Figure 3A that the views now in Third Angle Orthographic Projection are *exactly the same* as in First Angle Orthographic Projection but that the relative positions of the Plan and Elevation are different.

When the aforementioned rules are applied to objects situated in the 2nd and 4th angles their projections will be superimposed and therefore First or Third Angle Projection is obviously the most useful means of specifying the shape description of any object and both methods are used extensively.

Third Angle Projection is universal practice in the United States whilst First Angle Projection is most popular in Eastern Europe and the USSR. In the United Kingdom both modes of projection are in common use but it would appear that Third Angle Projection is gaining in popularity especially in the larger manufacturing industries.

However the current practice in most engineering drawing examinations is to supply details of the problem in First Angle Projection and permit the candidate to draw the solution in either First or Third Angle Projection. The solution to problems in this book are drawn in First or Third Angle Projection with no apparent preference for either mode, and the student should note that all views will be exactly the same regardless of the projection used and only the relative positions of the views will be different.

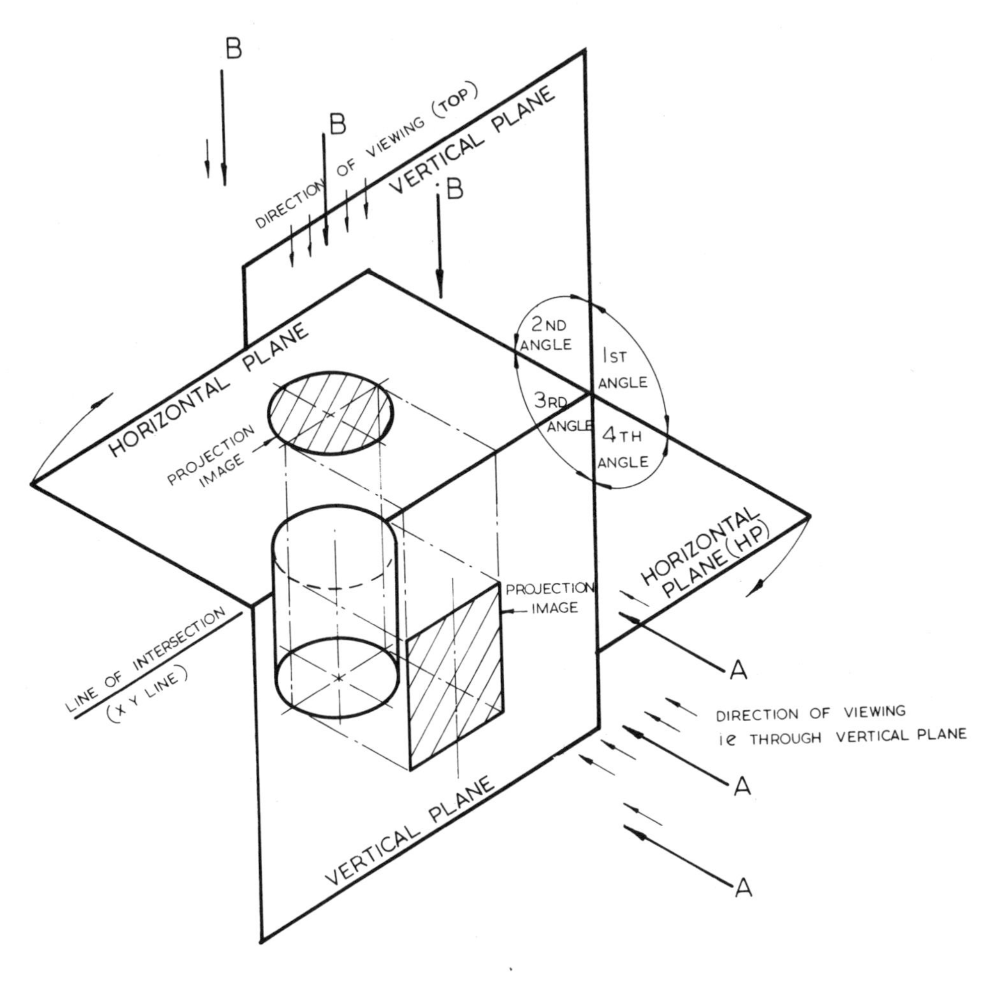

Figure 3B

Auxiliary Planes of Projection

The Elevation and Plan views may not always fully specify the shape description of an object. Consider the cylinder now positioned relative to the principal planes as shown in Figures 4 and 5. It can be seen that the Elevation and Plan views of a prism could be similar to that of the cylinder. Another view of the object is therefore necessary before the complete shape can be understood, and for this purpose another plane of projection termed an Auxiliary Vertical Plane (AVP) is introduced. This AVP is positioned perpendicular to both the VP and HP.

In First Angle Projection it was noted that the object is situated between the viewer and the plane of projection whilst in Third Angle Projection the plane of projection is placed between the object and the viewer.

This system is maintained when using auxiliary planes and the plane is also rabatted about XY until it is in the Vertical Plane.

Figures 4 and 5 illustrate the principle and it should again be noted that only the relative position of the views alters depending on whether First or Third Angle Projection is used.

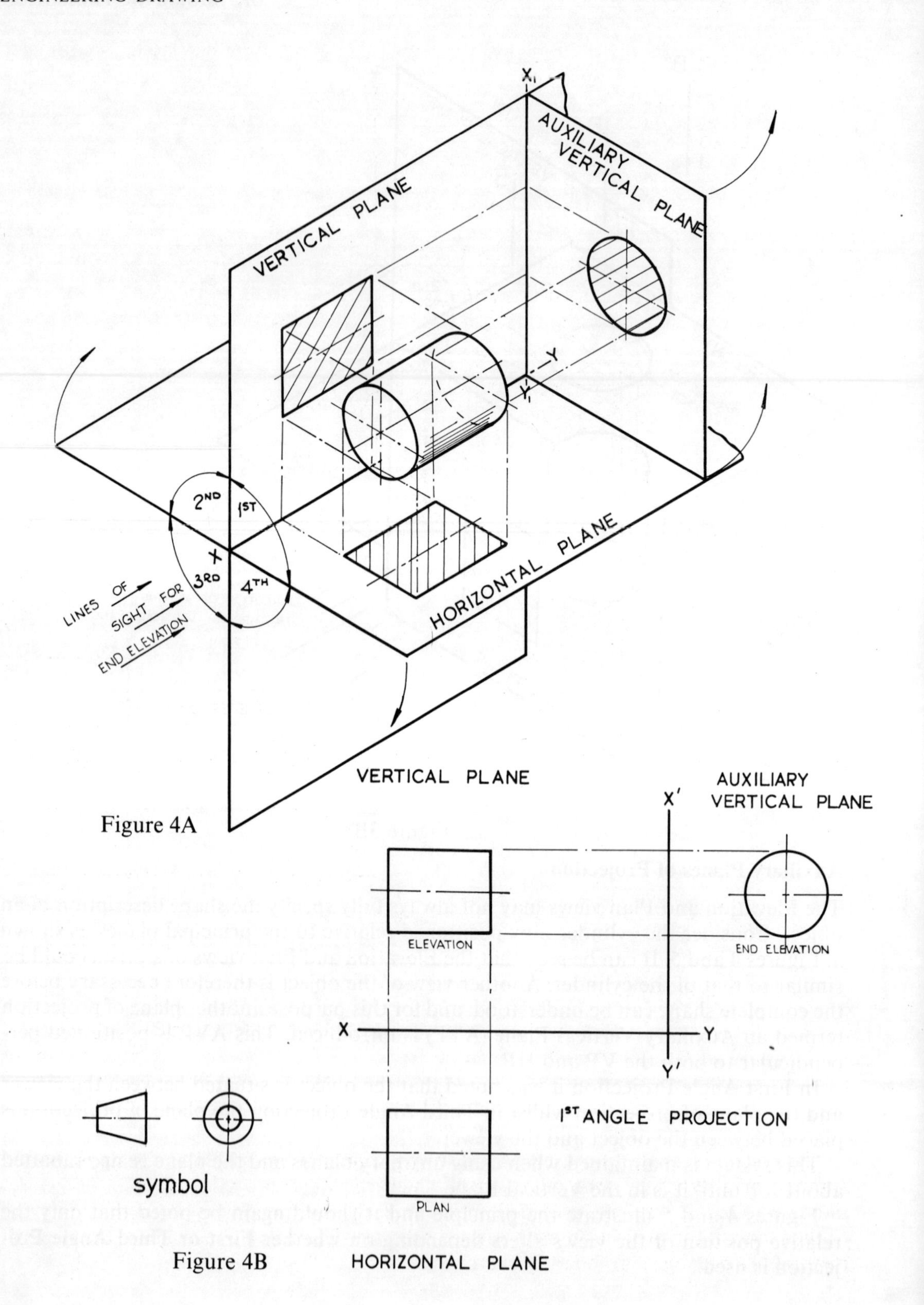

Figure 4A

Figure 4B

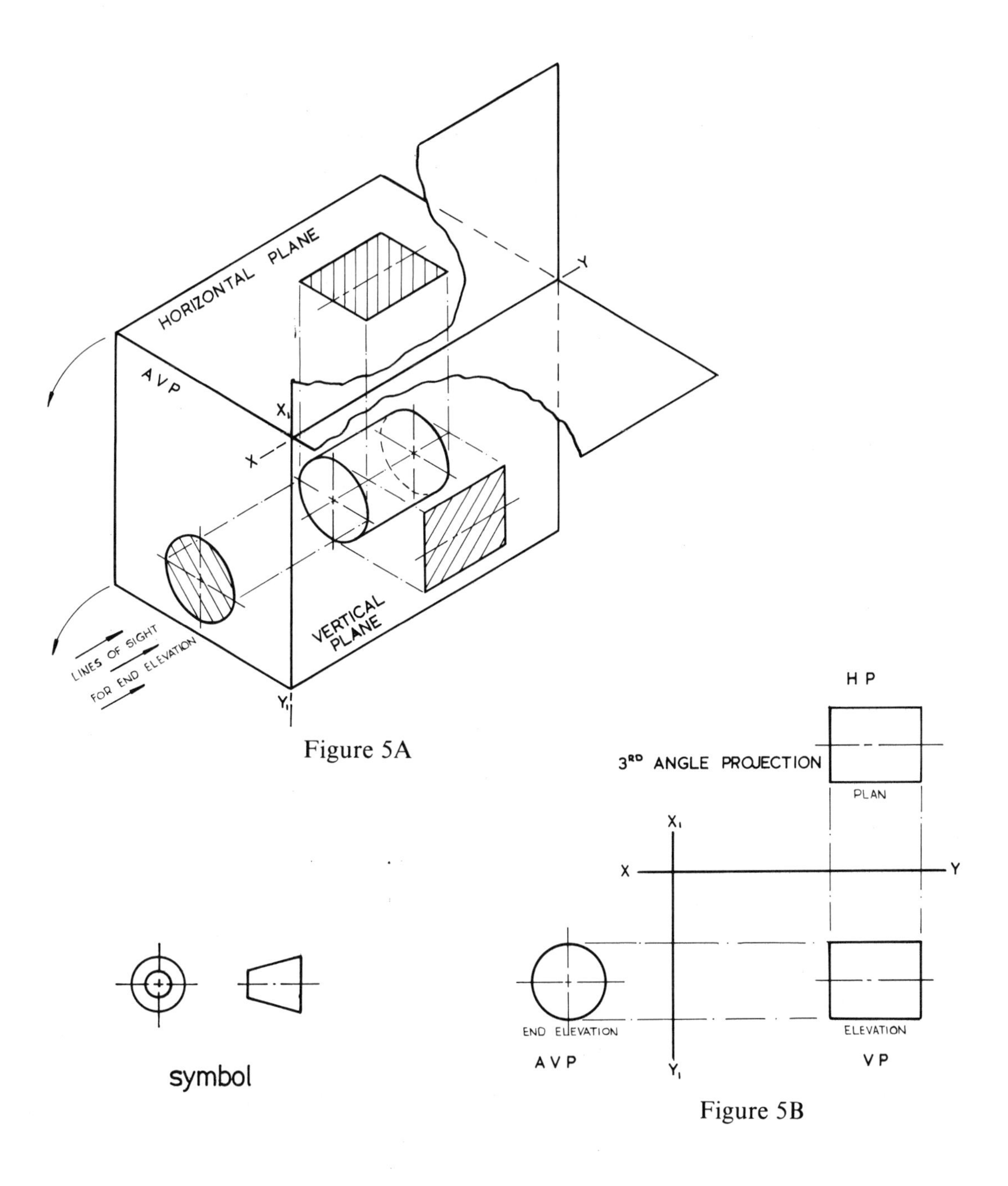

Figure 5A

Figure 5B

The system of projection used on a drawing should be indicated by the appropriate symbol, or alternatively the direction in which the views are taken should be clearly indicated. On the examples included in this book the author felt that it might be helpful to the reader if the projection method used was stated, and as a consequence the symbol method has been used in the early exercises only.

Projections on AVP that are perpendicular to both principal planes are commonly known as End Elevations. However it should be stated that terminology can often lead to confusion and it is advisable to consider each view simply as a projection on a plane, the views being positioned relative to each other by the direction of viewing and dependent on the type of orthographic projection used.

In examinations it is usual practice to state the view required by indicating the direction of viewing by an arrow. This method avoids the use of terms such as *inverted plan view* and *end plan view* when the required views may not comply with what the student may consider is the normal position for such a view.

It may often be necessary, in order to define the true shape description of an object to use a plane of projection that is not perpendicular to both the principal planes of reference. For example, the object illustrated in Figure 6 will not have the true shape of the surface ABCD revealed in any of the views so far considered.

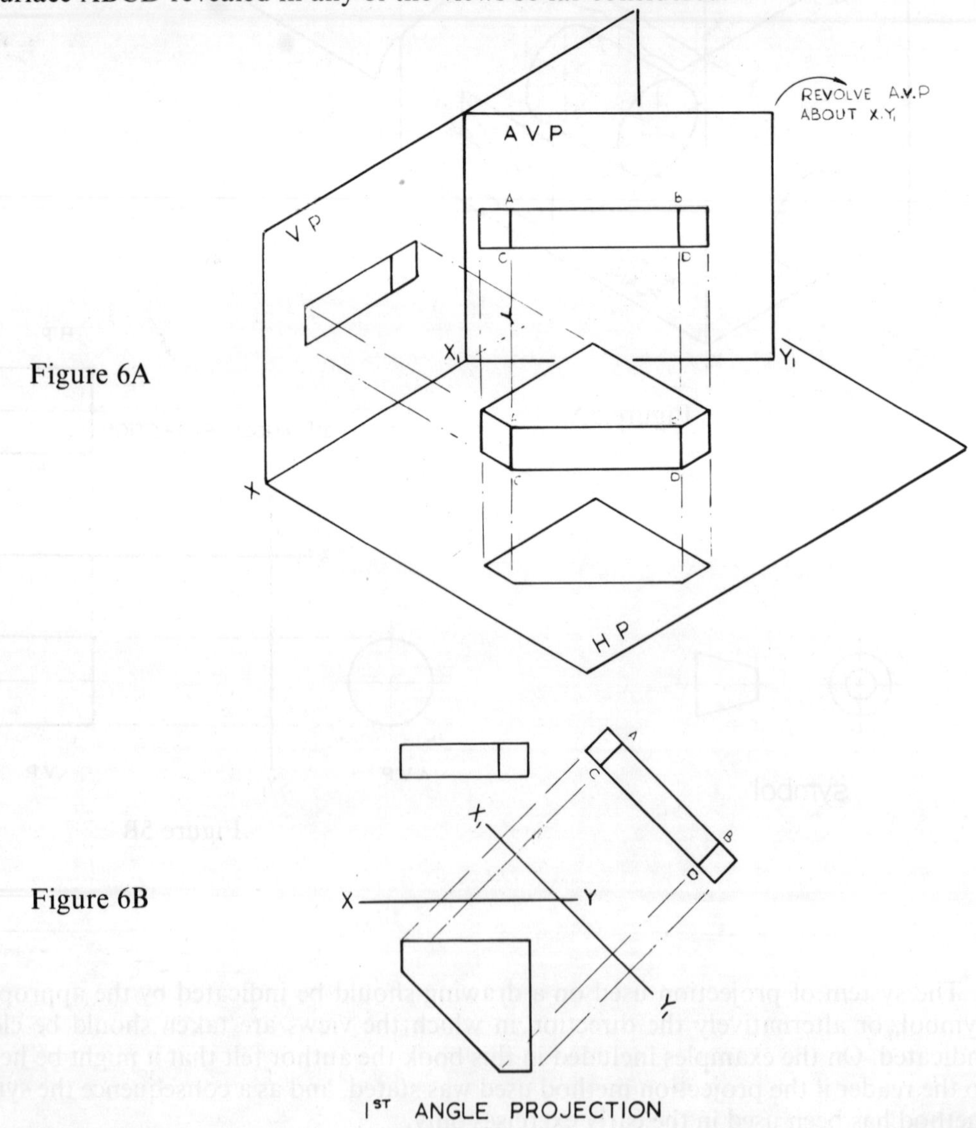

Figure 6A

Figure 6B

The true shape ABCD will be seen in a plane of projection that is parallel to this surface. In the example illustrated this plane of projection is still an AVP by definition since it is perpendicular to the HP although inclined to the VP. The auxiliary plane is now revolved towards the HP to obtain the orthographic projection.

A plane that is perpendicular to the VP and inclined at any angle other than 90° to HP is known as an Auxiliary Inclined Plane and may be used in a similar fashion to define the true shape of plane surfaces that are not parallel or perpendicular to the principal planes of reference.

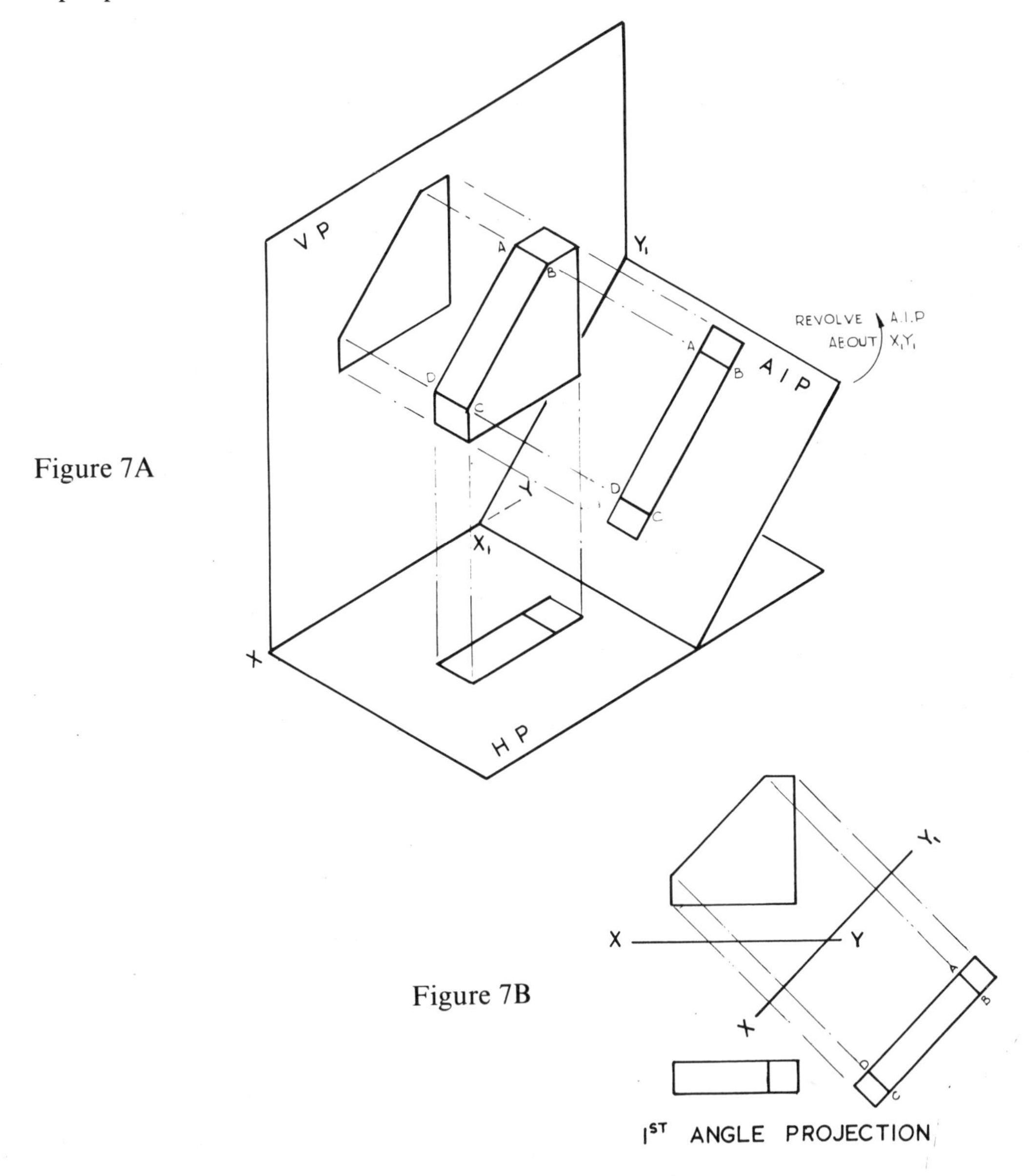

Figure 7A

Figure 7B

Figure 7 illustrates this example and it should be noted that the AIP is rabatted towards the VP in this instance.

4 SECTIONAL VIEWS (Single Components)

Except in the case of very elementary components it is likely that at least one sectional view will be helpful, and in some instances necessary, for complete shape description. A study of the examination papers in this book will reveal that in every paper a sectional view is involved, either in the presentation of the question or as a part of the required solution. This is not unusual since part of the examiner's task is to examine the candidates visualisation ability and sectional views do this very well.

In this chapter can be found an explanation of the technique involved together with the conventions as recommended in BS 308 and illustrations of the various types of sectional views.

Section Plane

In orthographic projection hidden details of a component are represented by broken lines, i.e. type 'D' lines (BS 308). Where these hidden features are numerous, overlapping or in anyway confusing, then sectional views are drawn. A cutting plane is selected which will enable a complete visualisation of the component. Figure 8 shows pictorially the application of such a plane.

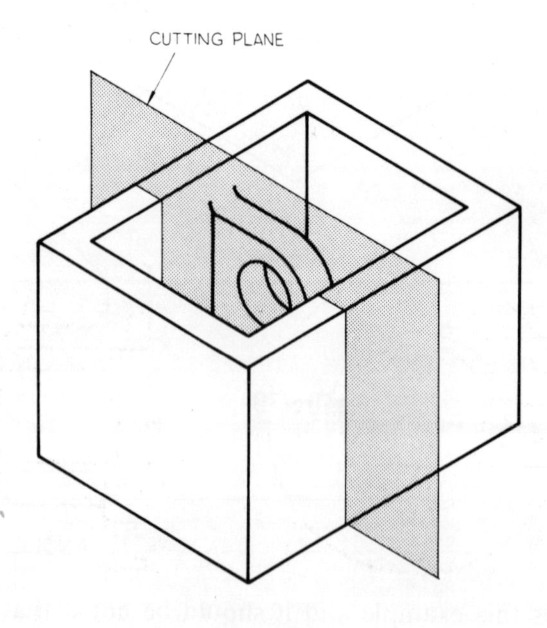

Figure 8

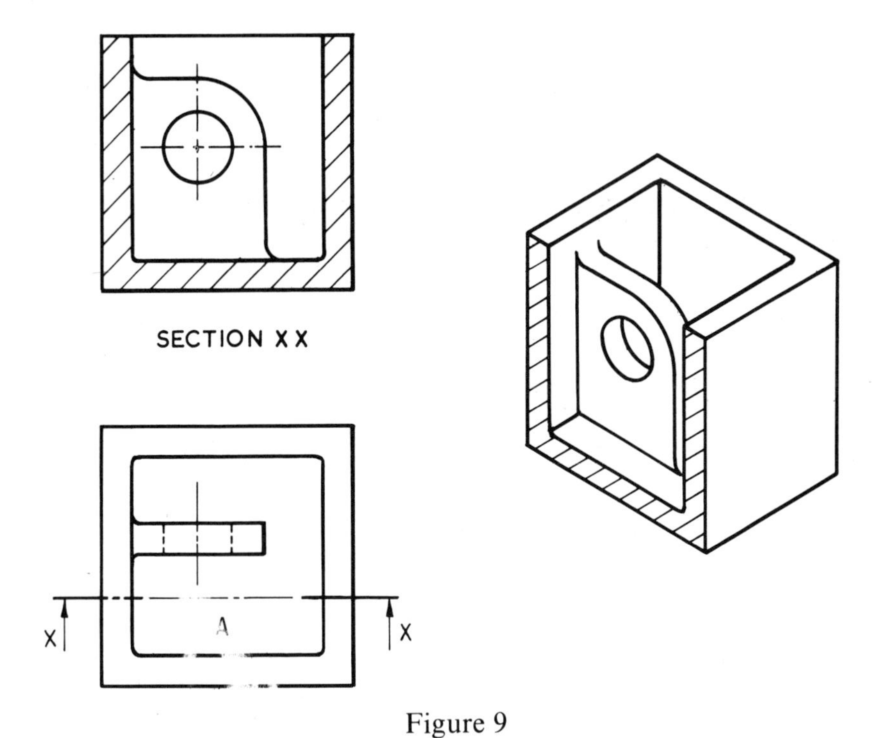

Figure 9

Figure 9 shows two orthographic views of the same component. An edge view of the cutting plane, represented by type 'F' lines (BS 308), has been shown in view A. Note that the section plane line has been labelled at each end with letters, i.e. 'XX', and the resulting sectional view entitled 'Section XX'.

When, as in Figure 10, the section is in one plane and along a centre line, it is not necessary to show a section plane line or section title.

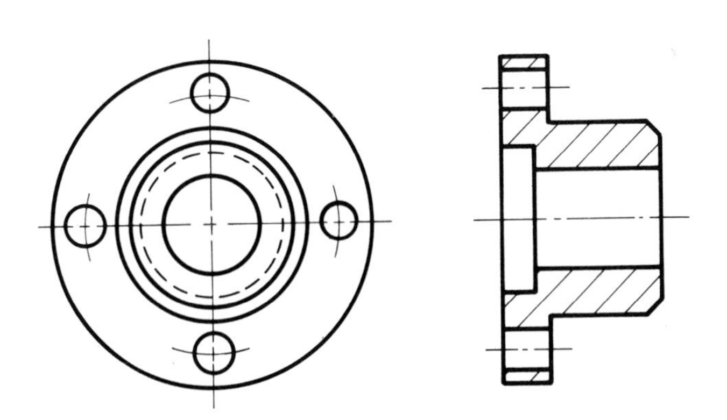

Figure 10

Cross-Hatching or Section Lining

A sectional view resulting from cutting through a component will show not only the outline of the cut material but also details of the component that are visible behind the cutting plane. In order that the shape description of the material actually cut is easier to visualise, symbolic lines are drawn on the cut surface. This technique, called cross-hatching or section lining, requires that type 'B' lines (BS 308) are drawn spaced from 3 mm to 6 mm apart depending on the area to be crosshatched. The section lines should normally be drawn at 45° to the edges of the drawing sheet. If the shape or position of the section would bring 45° sectioning parallel or nearly parallel to one of the sides, another angle may be chosen. Figure 11 shows an example of one of the various treatments possible when an angle of 45° cannot be used.

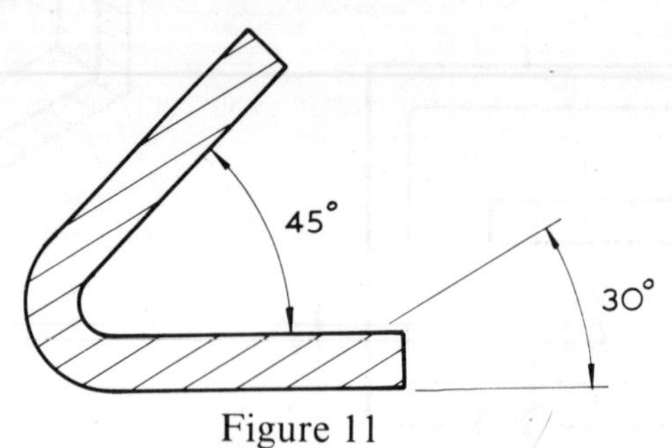

Figure 11

Hidden details are not normally shown on section views unless to clarify a feature whose shape description has not been conveyed precisely in other views.

Half Section

This method of sectioning enables the shape description of the outside and inside of a component to be shown in the minimum number of views (Figure 12). Normally only components which are symmetrical about a centre line are drawn in this way. Hidden lines on the half drawn as an outside view are shown only if necessary for dimensioning or clarity.

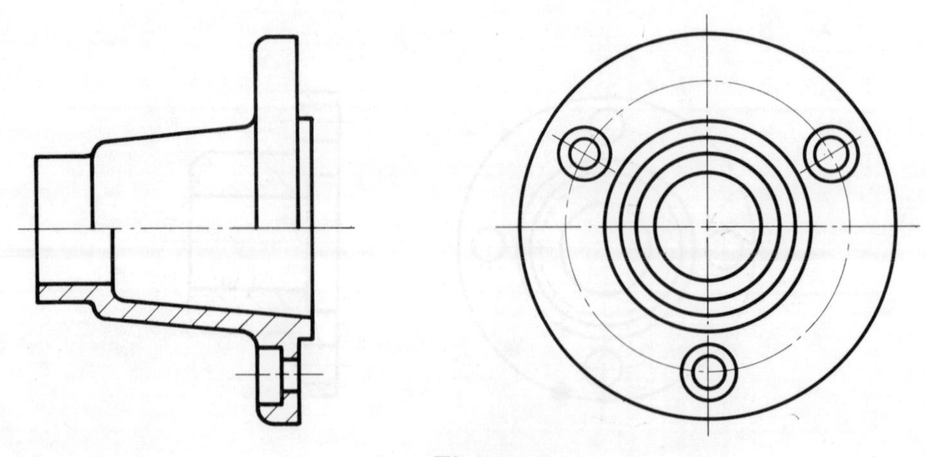

Figure 12

Part Section

When full sectioning of a component is more than is required for complete shape description, or in order to show a detail that would otherwise be hidden, a part section as shown in Figure 13 is drawn.

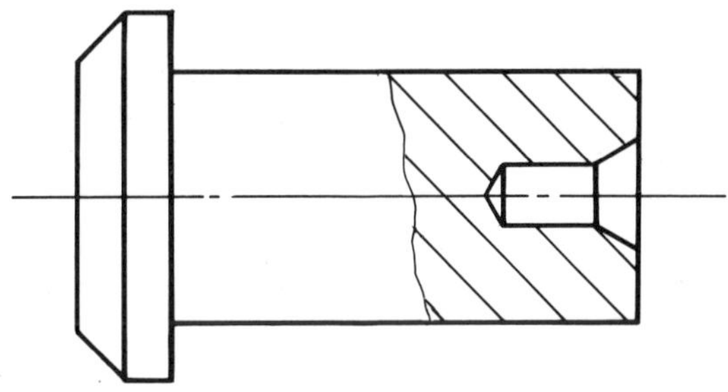

Figure 13

Revolved Section

Revolved sections enable the cross section of the component to be shown on an outside view. The cutting plane is revolved through 90° as shown in Figure 14.

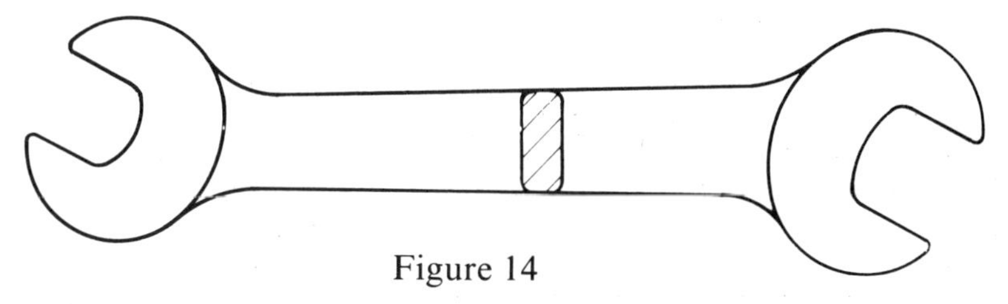

Figure 14

Offset Section

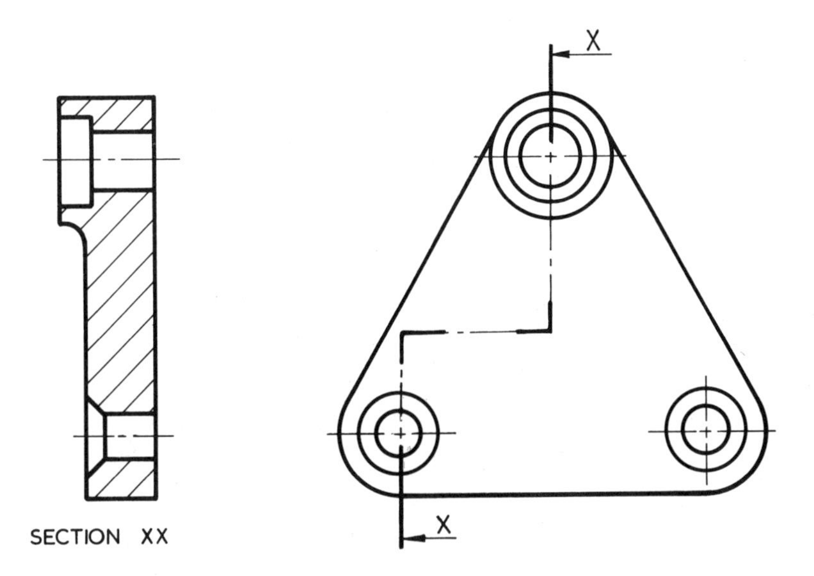

Figure 15A

The cutting plane is offset to include features of the component that could not be shown in section if a single cutting plane were used. See Figures 15A and 15B. N.B. Section AA is not in direct projection with the parent view and is known as a removed section.

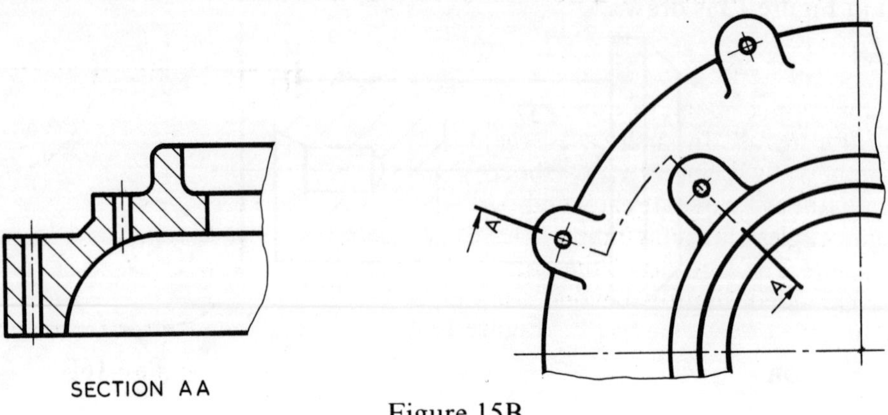

Figure 15B

Aligned Section

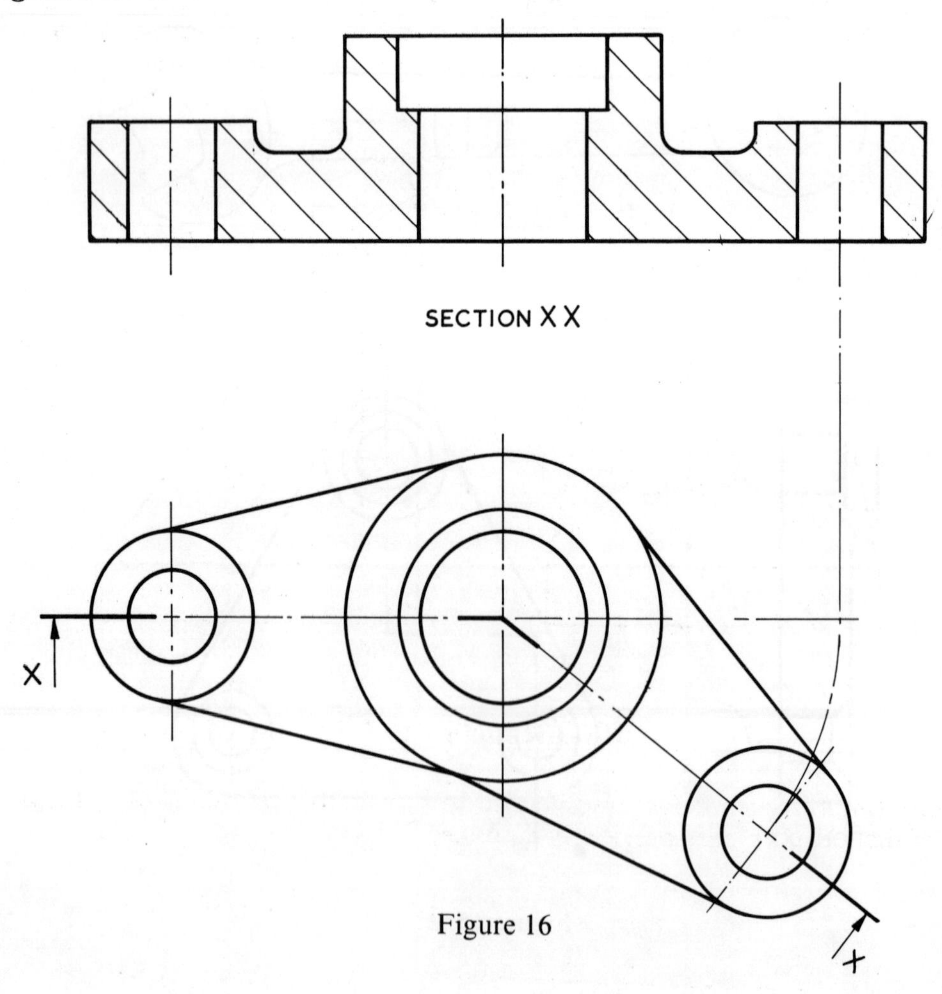

Figure 16

A conventional section along the horizontal centre line of the component shown in Figure 16 will provide details of only one of the two features. By rotating the cutting plane to a position which will pass through both features and then treating it as a single plane the complete shape description can be shown in two views.

Symmetry

It is not always necessary to draw symmetrical parts in full. In such cases, the line of symmetry is defined by two short thick parallel lines drawn at each end of and at right angles to the symmetry demarcation line. The outline of the component is extended slightly beyond the line of symmetry.

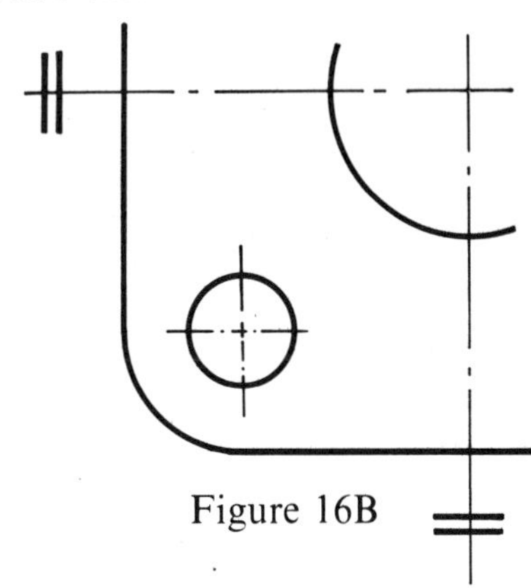

Figure 16B

Exceptions

Ribs. The reinforcing ribs or webs of a casting are not cross hatched when the cutting plane passes through them longitudinally. This treatment has been illustrated in Figure 17. The rib shown is relatively thin and to cross-hatch it would give a false sense of robustness to the component and would also serve no useful purpose.

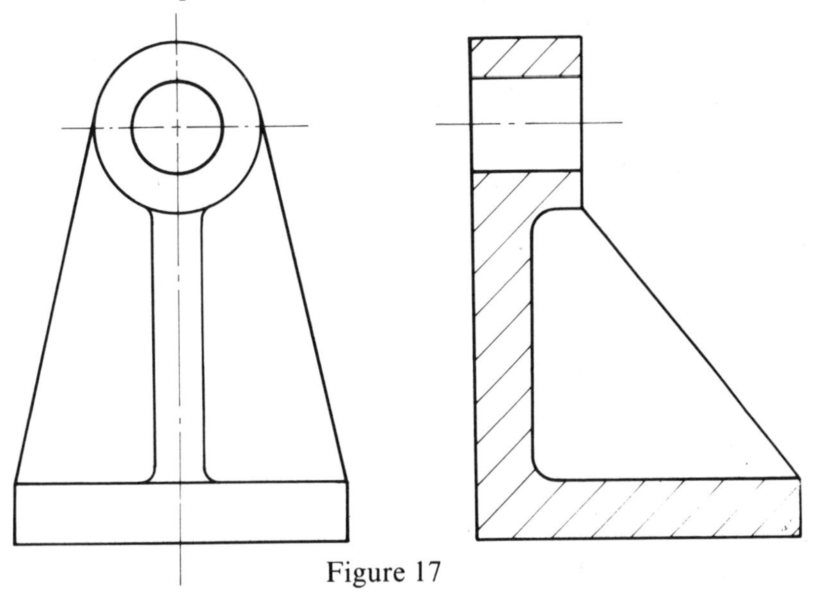

Figure 17

PROBLEMS 1 to 3

The following three problems are intended to provide typical examples of the use of some of the section planes previously described.

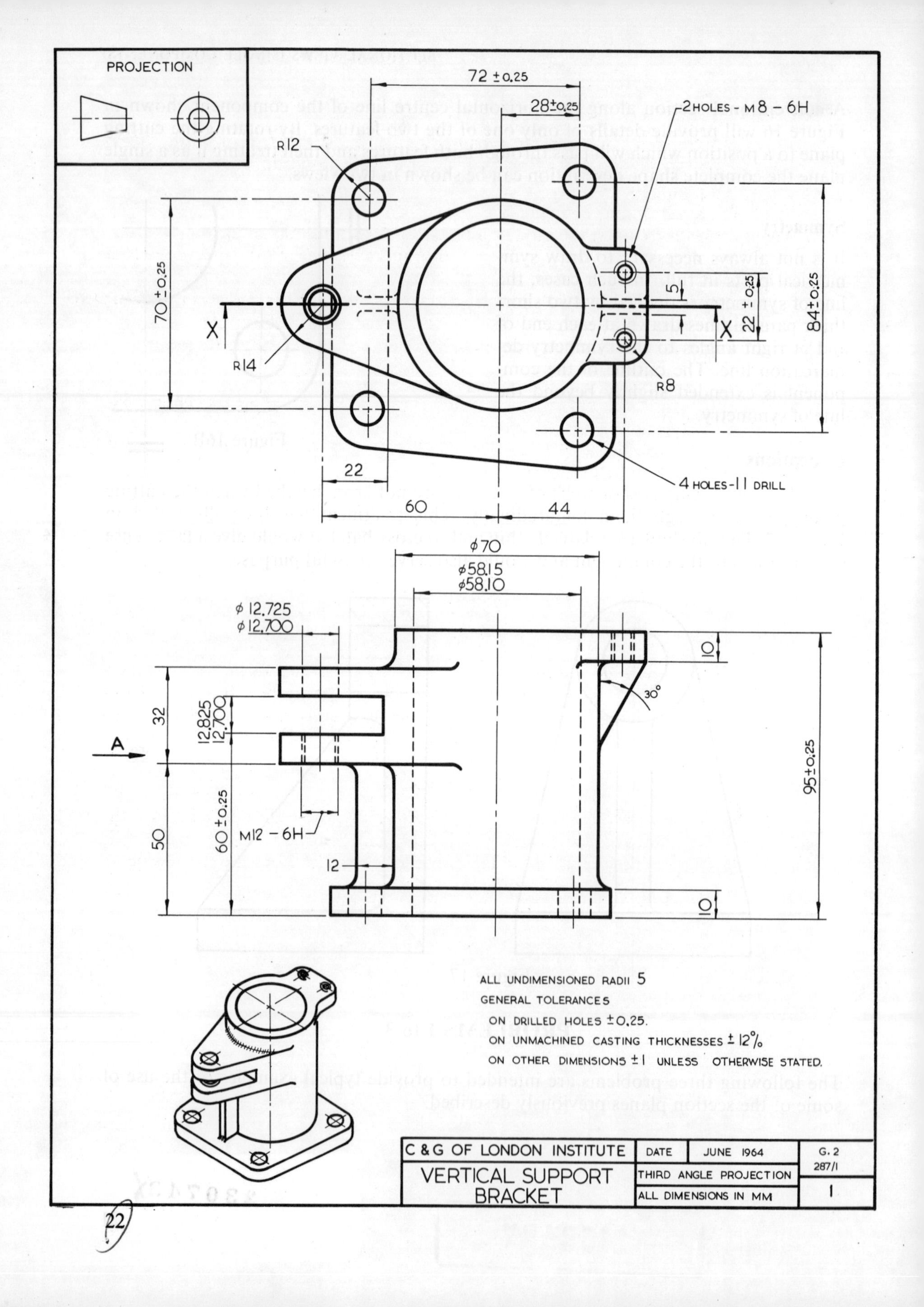
PROJECTION
72 ±0.25
28±0.25
2 HOLES - M8 - 6H
R12
70 ±0.25
84±0.25
22 ±0.25
5
X
R14
R8
22
4 HOLES-11 DRILL
60
44
ϕ70
ϕ58.15
ϕ58.10
ϕ 12.725
ϕ 12.700
10
30°
32
12.825
12.700
A
95±0.25
50
60 ±0.25
M12 - 6H
12
10
ALL UNDIMENSIONED RADII 5
GENERAL TOLERANCES
ON DRILLED HOLES ±0.25.
ON UNMACHINED CASTING THICKNESSES ± 12%
ON OTHER DIMENSIONS ±1 UNLESS OTHERWISE STATED.
C & G OF LONDON INSTITUTE
DATE
JUNE 1964
G.2
287/1
VERTICAL SUPPORT
BRACKET
THIRD ANGLE PROJECTION
ALL DIMENSIONS IN MM
1

Problem 1

Figure 1 shows two views of a **vertical support bracket** in 3rd angle projection. Draw the following views full size using 3rd or 1st angle projection and BSI conventions:

(a) The given plan.
(b) Sectional elevation on the plane XX.
(c) An elevation looking in the direction of arrow A without hidden detail.

Add a title and scale, and the dimensions and notes necessary for drilling the four holes in the base.

Show machining symbols on the upper and lower faces of the bracket and inside the 58·15 mm bore

State which projection angle you have used.

Time allowed – $1\frac{1}{4}$ hours.

Solution (page 131)

The pictorial view on the question paper has been added as an aid to visualisation and was not included in the question as set.

View (a)

The two given views have been drawn in third angle projection where the normal plan view is positioned above the elevation. Consequently after deciding on a projection angle a direct copy can be made of the top given view.

View (b)

The required section view has a similar profile to that of the given outside elevation and, depending on the angle of projection chosen, should be positioned above or below view (a). It should be noticed that the section plane passes through two webs but in a manner which requires that they are not cross-hatched in the sectional view.

View (c)

This view should be projected from view (b) and to its left or right depending on the angle of projection chosen. The 12·825/12·700 mm space between the two flanges will appear as a rectangle, the length of which can be determined from the hidden detail line in view (a).

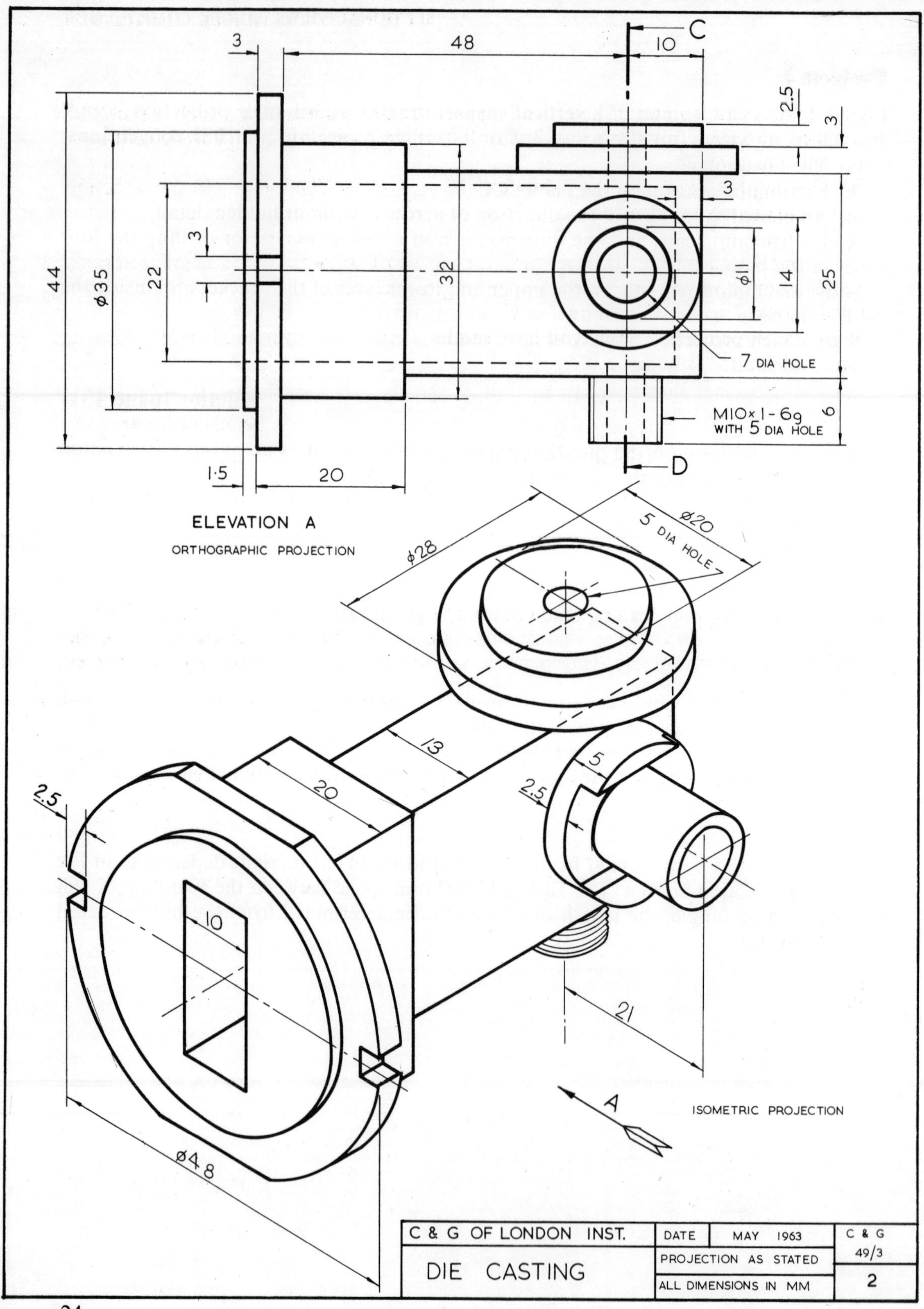
3
48
C
10
2.5
3
3
3
44
⌀35
22
32
⌀11
14
25
7 DIA HOLE
M10×1-6g
WITH 5 DIA HOLE
6
1·5
20
D
ELEVATION A
ORTHOGRAPHIC PROJECTION
⌀28
⌀20
5 DIA HOLE
13
5
20
2.5
2.5
10
21
A
ISOMETRIC PROJECTION
⌀48
C & G OF LONDON INST.
DATE
MAY 1963
C & G
49/3
PROJECTION AS STATED
DIE CASTING
ALL DIMENSIONS IN MM
2

Problem 2

Figure 2 shows an isometric sketch and an elevation of a **die casting** which is part of a waveguide. Do not copy these views, but draw three views, twice full size as follows:

(a) The elevation A.
(b) The sectional end elevation, showing the section on CD, looking in the direction of the arrows at C and D.
(c) The plan.

All three views are to be in either first angle or third angle projection, and you must state which method you have used.

Hidden edges need not be shown in views (b) and (c).

Show the three maximum overall dimensions – length, breadth and height.

Print, neatly, the title and scale.

Time allowed – $1\frac{2}{3}$ hours.

Solution (page 132)

View (a)

Having chosen the method of projection, and hence the orientation of the required views, the given elevation can be copied.

View (b)

The section plane cuts through the centre of the top diameters, through the hollow rectangular section and the centre of the tubular leg and then through the centre of the threaded spigot. Behind the section plane will be seen the end plate and the larger rectangular section.

View (c)

The shapes described above are now seen in plan view. The length of the flats on the left hand end plate can be determined from view (b).

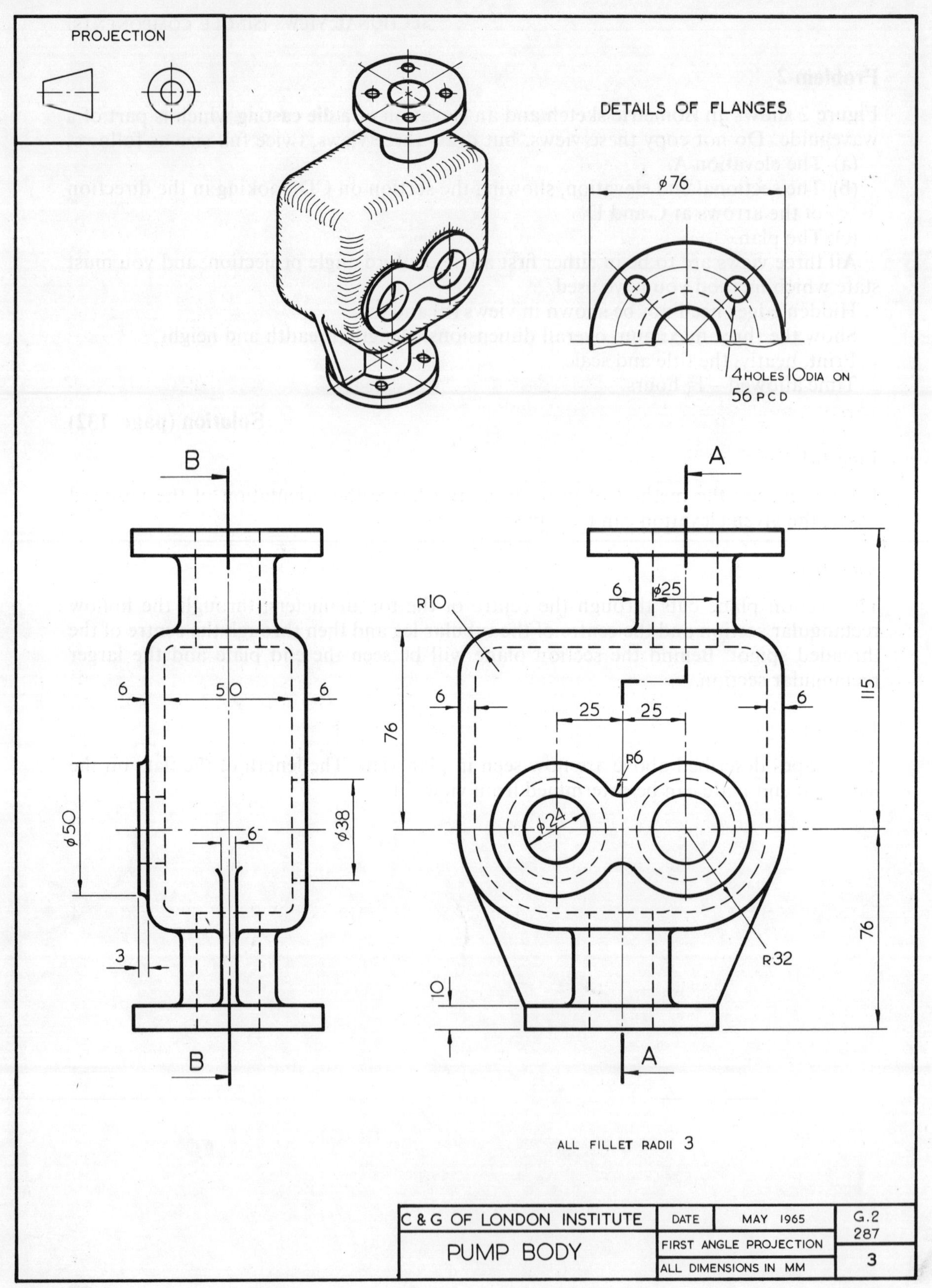
PROJECTION
DETAILS OF FLANGES
ϕ76
4 HOLES 10 DIA ON
56 PCD
B
A
6
ϕ25
R10
6
50
6
6
25
25
6
76
115
R6
ϕ24
ϕ50
ϕ38
6
76
R32
3
10
B
A
ALL FILLET RADII 3
C & G OF LONDON INSTITUTE
DATE
MAY 1965
G.2
287
PUMP BODY
FIRST ANGLE PROJECTION
ALL DIMENSIONS IN MM
3

Problem 3

Figure 3 shows, in first angle projection, details of a **pump body**.

Do not copy these views but draw full size in first or third angle projection:

(a) A sectional elevation AA.

(b) A sectional elevation BB.

(c) A plan projected from view (b).

Hidden details should be shown in view (c) only. Add six principal dimensions, a title and scale, and state clearly the method of projection you have used.

Time allowed – $1\frac{2}{3}$ hours.

Solution (page 133)

The pictorial view on the question paper has been added as an aid to visualisation and was not included in the question as set.

View (a)

This is an example of a component cut by an offset section plane and results in both flanges being cut through their centres. The viewing position for Section AA is similar to the given end elevation. Consequently, the required sectional view can be built up from the profile of the given end elevation and changing hidden detail lines to full lines. The distance marked P on Section AA can be projected from view (b) as shown.

View (b)

Depending on the angle of projection chosen this view will be positioned to the left or right of Section AA. The viewing position for Section BB is similar to that for the given front elevation and thus Section BB has a similar profile.

It is necessary to construct lightly a portion of the outline of the 3 mm thick boss in order that the width cut by the section plane AA can be projected into view (a).

The two 6 mm thick webs are cut by the section plane but in a manner which requires that they are not cross-hatched.

View (c)

Depending on the angle of projection chosen, this view will be positioned above or below view (b). Hidden detail is required in this view. N.B. the portion of the flange shown in the question paper giving details of the 10 mm dia holes is called an auxiliary view and, as is the custom, has been drawn in third angle projection.

5 MACHINING OPERATIONS

A detail drawing will normally provide a complete shape description of the component and include all of the dimensions necessary for the component's manufacture. Quite often it is possible for a dimension to include a description of the tool used to produce the feature and it is not unusual for the shape description of a feature to be given in note form only, i.e., 6 mm c' bore, or spotface 25 mm dia. Consequently it is important that the student becomes familiar with the names and uses of the more common tools.

Centre Drill

A centre drill is used for locating a hole prior to drilling.

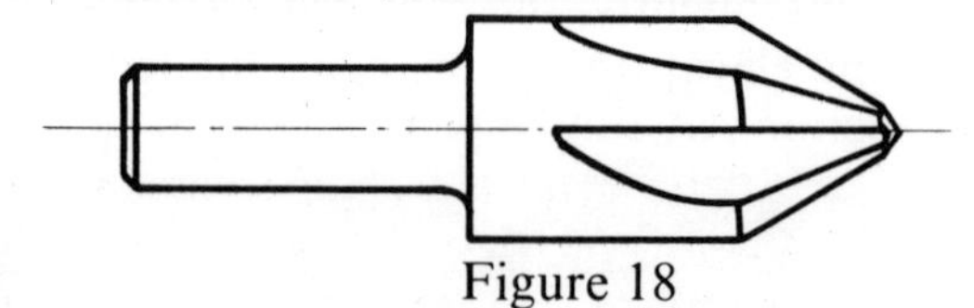

Figure 18

Twist Drill

This tool is used for producing holes where a high degree of accuracy is not required, i.e., clearance holes for bolts, etc.

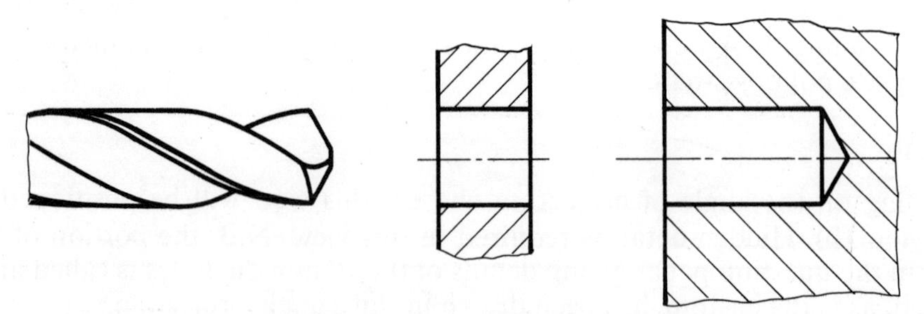

Figure 19

Reamer

A reamer is used to improve the accuracy and surface finish of a drilled hole.

Figure 20

Countersink and Counterbore

Where it is necessary to enlarge a hole for the acceptance of a bolt head a countersink (Figure 21) or counterbore (Figure 22) is used.

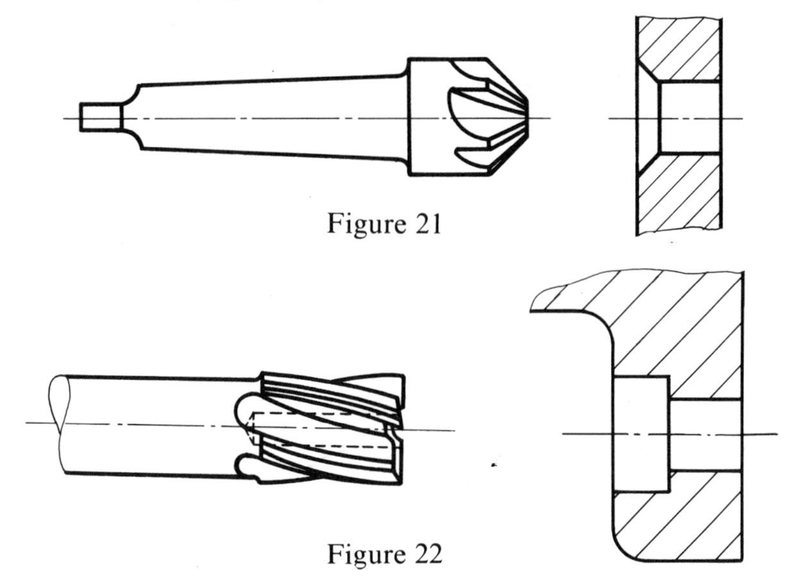

Figure 21

Figure 22

Spotfacing is the name given to the process of removing only a small amount of metal, using thc counterborc, in ordcr to provide a smooth surface finish for the seating of a bolt head. The surface to be machined may be in the form of a raised circular 'boss' as in Figure 23A or, as shown in Figure 23B, in the form of a shallow counterbore.

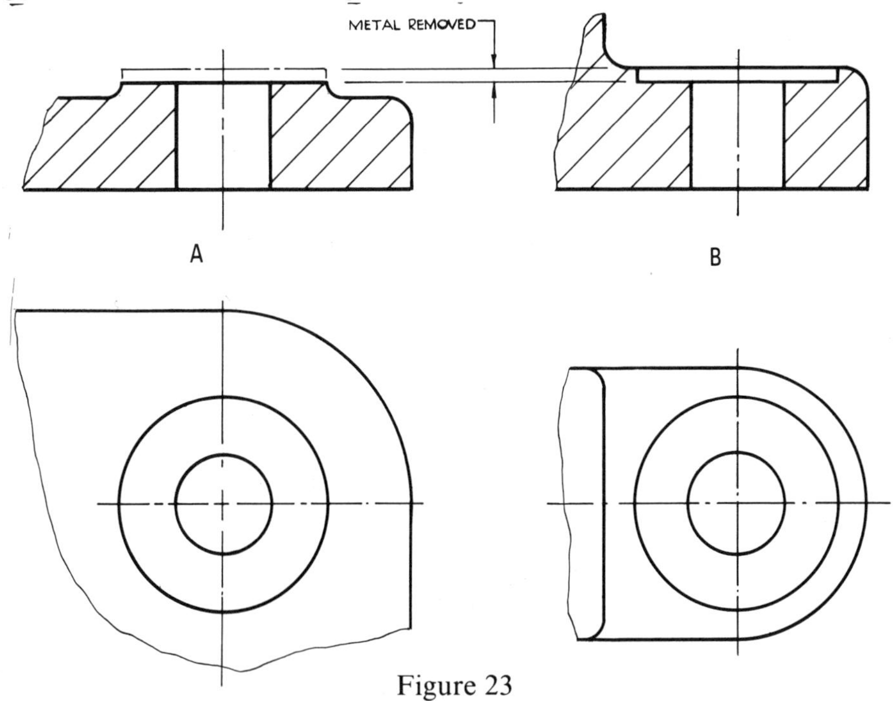

Figure 23

Tap and Die

Standard threads, internal and external, are cut by taps (Figure 24) and dies (Figure 25) respectively.

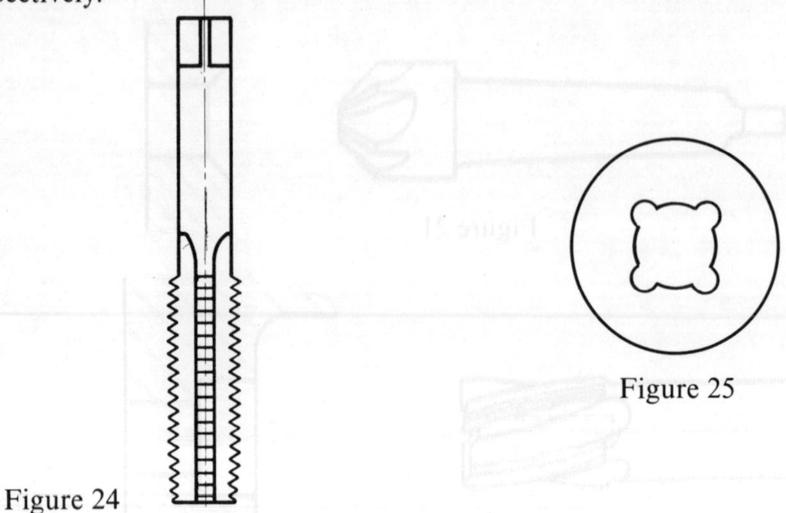

Figure 24

Figure 25

A threaded hole is obtained by first drilling a hole of a diameter approximately the size of the core diameter of the thread, i.e., the diameter across the bottom of the thread groove.

Conversely an external thread is cut on a spindle whose diameter is approximately the maximum diameter of the thread, i.e., the size by which the thread is known e.g., M10.

Because of the build up of swarf at the bottom it is necessary when threading a blind hole to drill deeper than the required depth of thread.

Similarly, in order that there is sufficient full depth thread (the end of a tap is tapered) it is necessary to tap to a greater depth than the envisaged penetration by the screw or bolt.

Figure 26 illustrates the conditions necessary for a bolt which is to be screwed in a depth of 20 mm.

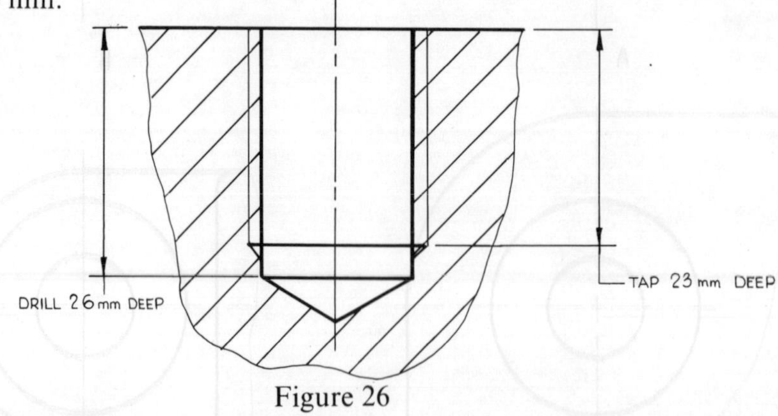

Figure 26

N.B. The threaded hole above has been drawn in accordance with the convention shown in BS 308.

PROBLEMS 4 to 13

There now follow ten problems of increasing difficulty which further illustrate the various section planes available, and also examples of the machining operations previously described.

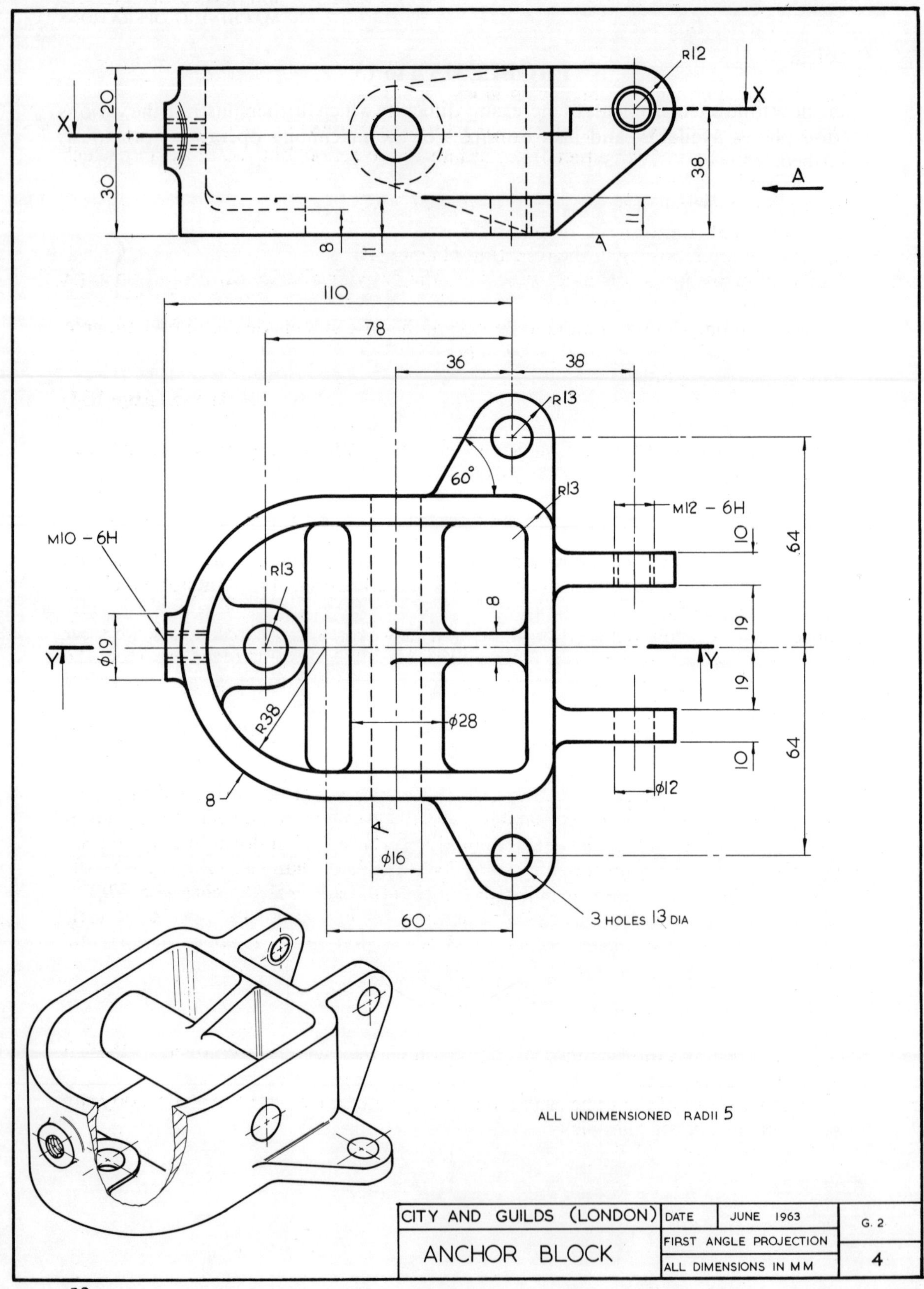

R12
X
X
20
30
38
A
8
11
110
78
36
38
R13
60°
R13
M12 – 6H
M10 – 6H
R13
10
64
8
19
Y
Y
ϕ19
19
R38
ϕ28
64
10
8
ϕ12
A
ϕ16
60
3 HOLES 13 DIA
ALL UNDIMENSIONED RADII 5
CITY AND GUILDS (LONDON)
DATE
JUNE 1963
G. 2
FIRST ANGLE PROJECTION
ANCHOR BLOCK
ALL DIMENSIONS IN MM
4

Problem 4

Figure 4 shows two views of an **anchor block**.

Draw the following views full size, using BSI conventions.

Your drawings may be in either 1st or 3rd angle projection, but you must state which you have used.

(a) A plan, the top half of which is a section on XX.
(b) A sectional elevation on the plane YY.
(c) An elevation looking in the direction of arrow A.

Add six dimensions, a title and the scale to the drawing. State what projection angle has been used.

Show machining symbols on the underside of the base and inside the 16 mm dia hole.

Hidden detail is required on view (a) only.

Time allowed – $1\frac{1}{4}$ hours.

Solution (page 134)

The pictorial view on the question paper has been added as an aid to visualisation and was not included in the question as set.

View (a)

The lower half of this view is a direct copy of the given plan view. The top half, a section on the offset cutting plane XX, will contain all of the features of the Anchor Block but most will be in section.

Those features below the section plane, and hence not cross hatched, are the three clamping lugs or feet at the base. Hidden detail has been asked for in this view. To avoid repetition of given information, the solution shows this view as a complete section.

View (b)

Depending on the angle of projection chosen, this view will be positioned above or below view (a). The required sectional view will have a similar profile to the given elevation and can be built up by drawing full lines where hidden detail lines have been shown. The 8 mm thick web supporting the 28 mm outside diameter tube is cut by the section plane but in a manner which requires that it is not cross-hatched (BS 308).

Because the tapped hole is in the rear lug, the given elevation shows two concentric circles where as in view (b) the conventional representation for a tapped hole should be used.

View (c)

This view, depending on the angle of projection chosen, will be positioned to the left or right of view (b).

The merger point of the upper surface of the two outer feet with the main body can be determined from the plan view.

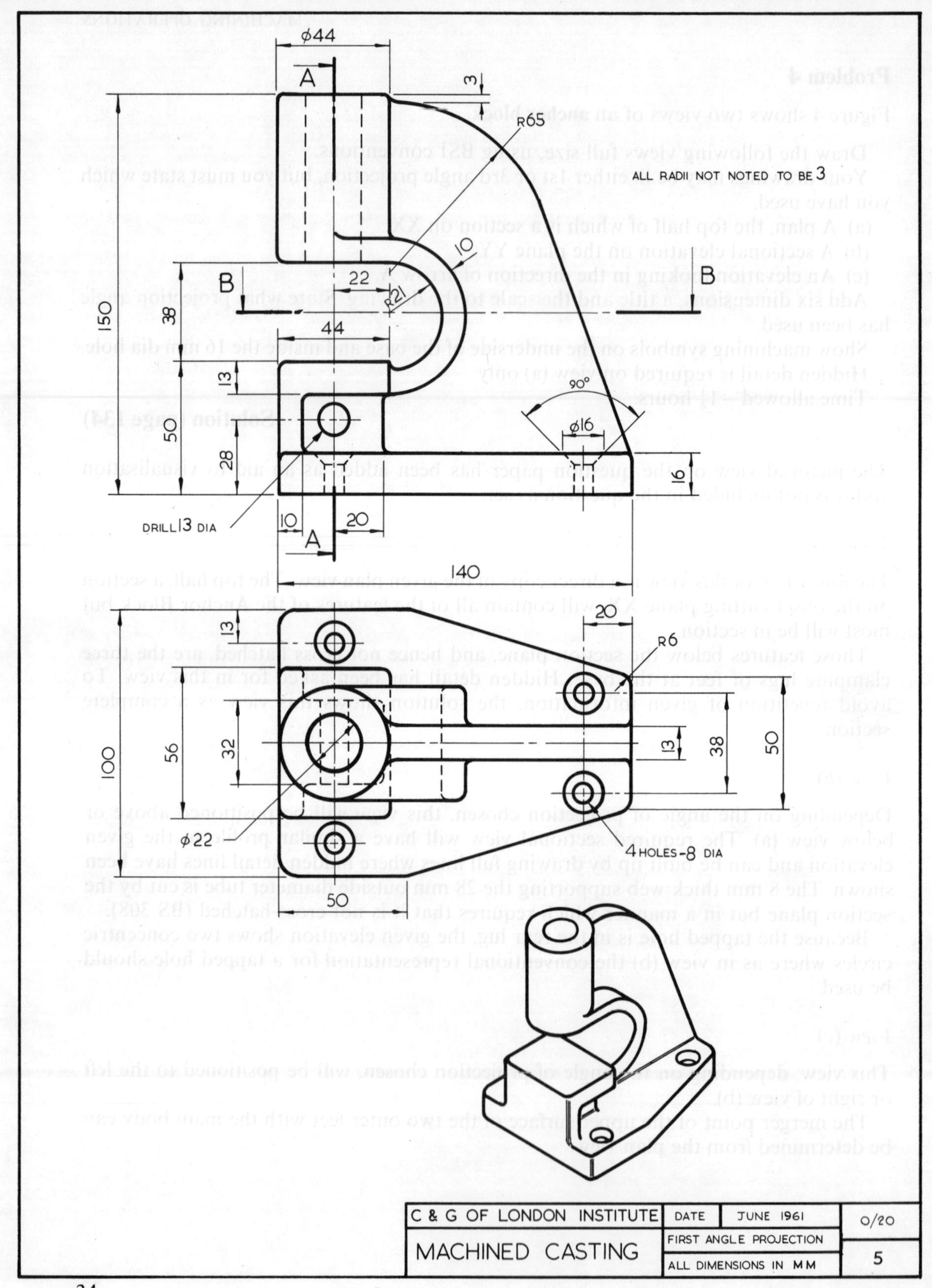
ϕ44
A
3
R65
ALL RADII NOT NOTED TO BE 3
10
B
22
R21
B
150
38
44
13
90°
50
ϕ16
28
16
DRILL 13 DIA
10
20
A
140
20
13
R6
100
56
32
13
38
50
ϕ22
4 HOLES-8 DIA
50
C & G OF LONDON INSTITUTE
DATE
JUNE 1961
0/20
MACHINED CASTING
FIRST ANGLE PROJECTION
ALL DIMENSIONS IN MM
5

Problem 5

Figure 5 shows the elevation and plan of a **machined casting**. Draw full size in either first or third angle projection, the following views:

(a) The elevation as shown.
(b) The sectional plan, the section being taken on the line BB.
(c) The sectional elevation taken on the line AA, looking in the direction of the arrows.

All hidden details must be included. Add SIX of the leading dimensions.
Time allowed $-1\frac{3}{4}$ hours.

Solution (page 135)

The pictorial view on the question paper has been added as an aid to visualisation and was not included in the question as set.

View (a)

The two given views have been drawn in first angle projection where the normal plan view is projected below the elevation. Therefore view (a) is a direct copy of the top given view.

View (b)

The viewing direction is similar to that of the given plan view and, depending on the angle of projection chosen this view will be projected above or below view (a). With a line representing the plane BB drawn in view (a) it can be established that the 44 mm diameter tube and part of its supporting section is above the section plane and hence not seen in view (b).

View (c)

Depending on the angle of projection chosen this view will be projected to the left or right of view (a). As an aid to visualisation the cutting plane AA should be identified in both of the given views. As shown in the pictorial view of section AA the plane cuts through: (1) the centre of the 44 mm outside diameter tube, (2) the platform of 'T' section on the centre of the 13 mm diameter hole, and (3) through the base flange on the centre of two countersunk holes.

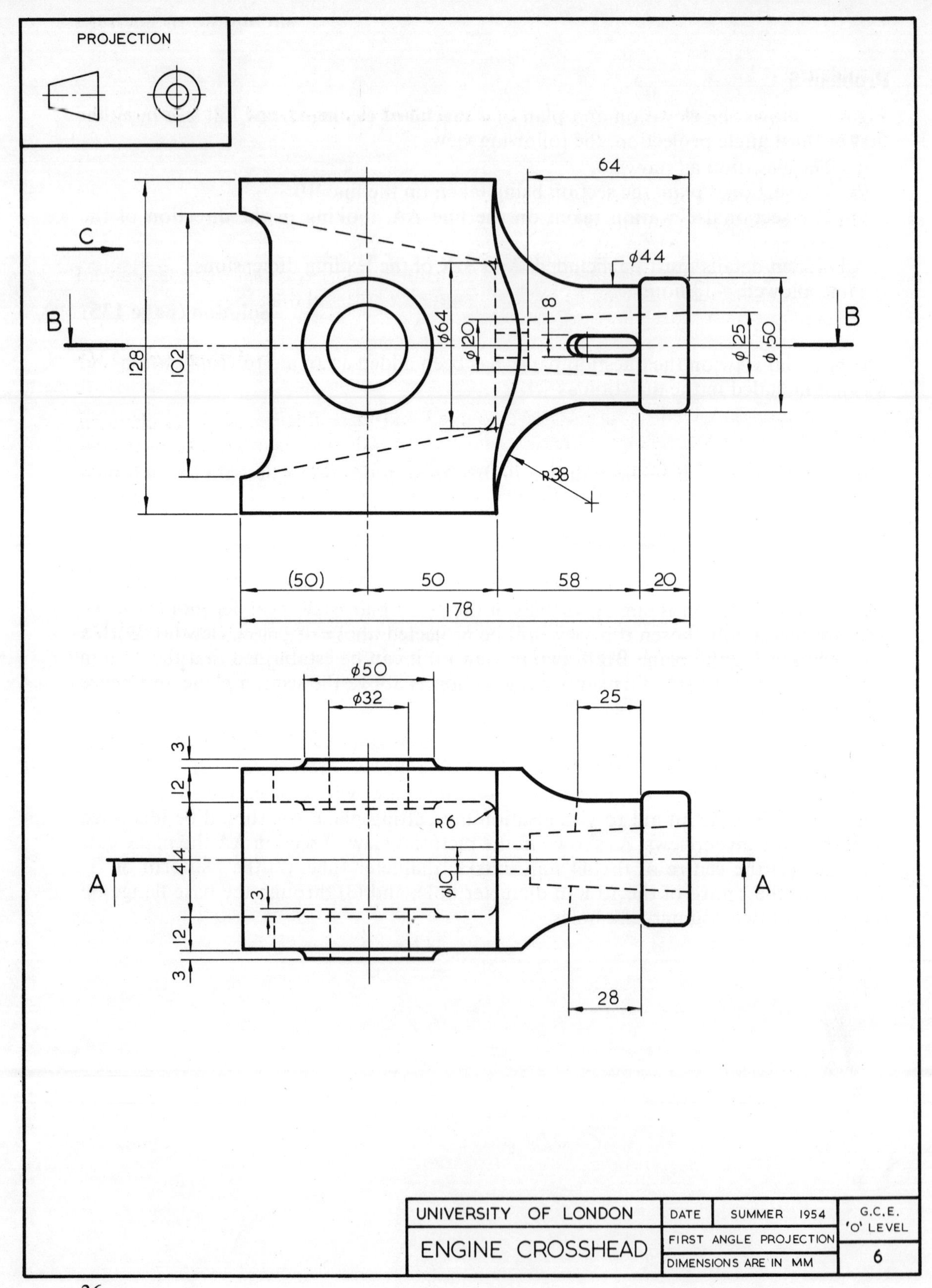
PROJECTION
C
B
B
64
ϕ44
8
ϕ64
ϕ20
ϕ25
ϕ50
128
102
R38
(50)
50
58
20
178
ϕ50
ϕ32
25
3
12
R6
44
ϕ10
A
A
3
12
3
28
UNIVERSITY OF LONDON
DATE
SUMMER 1954
G.C.E. 'O' LEVEL
ENGINE CROSSHEAD
FIRST ANGLE PROJECTION
DIMENSIONS ARE IN MM
6

Problem 6

Figure 6 gives an elevation and plan of part of an **engine crosshead**. Do not copy the given views.

Draw full size:
(a) A sectional front elevation on line AA.
(b) A sectional plan on line BB.
(c) An end elevation looking in the direction of the arrow C.
After completing your drawing add six leading dimensions, the title Engine Crosshead and the scale. Dotted lines may be omitted from views (a) and (b).
Time allowed – $1\frac{1}{2}$ to $1\frac{3}{4}$ hours.

Solution (page 136)

View (a)

The profile of the given elevation can be copied and the hidden detail lines changed to the full lines. The position for the two semicircular ends of the slot must be projected from view (b). Note the bracket used to indicate the dimensions to be used for reference only, i.e. (50), in accordance with BS 308.

View (b)

Again, the profile of the given plan can be copied and the hidden detail lines changed to full lines. Note that point B can be determined by projection as shown, but is the result of constructing an arc of 38 mm radius on a line projected vertically through the centre of the 38 mm radius in view (a).

View (c)

This view relies on complete visualisation from views (a) and (b) but consists of a rectangular frame and four 3 mm thick bosses in elevation.

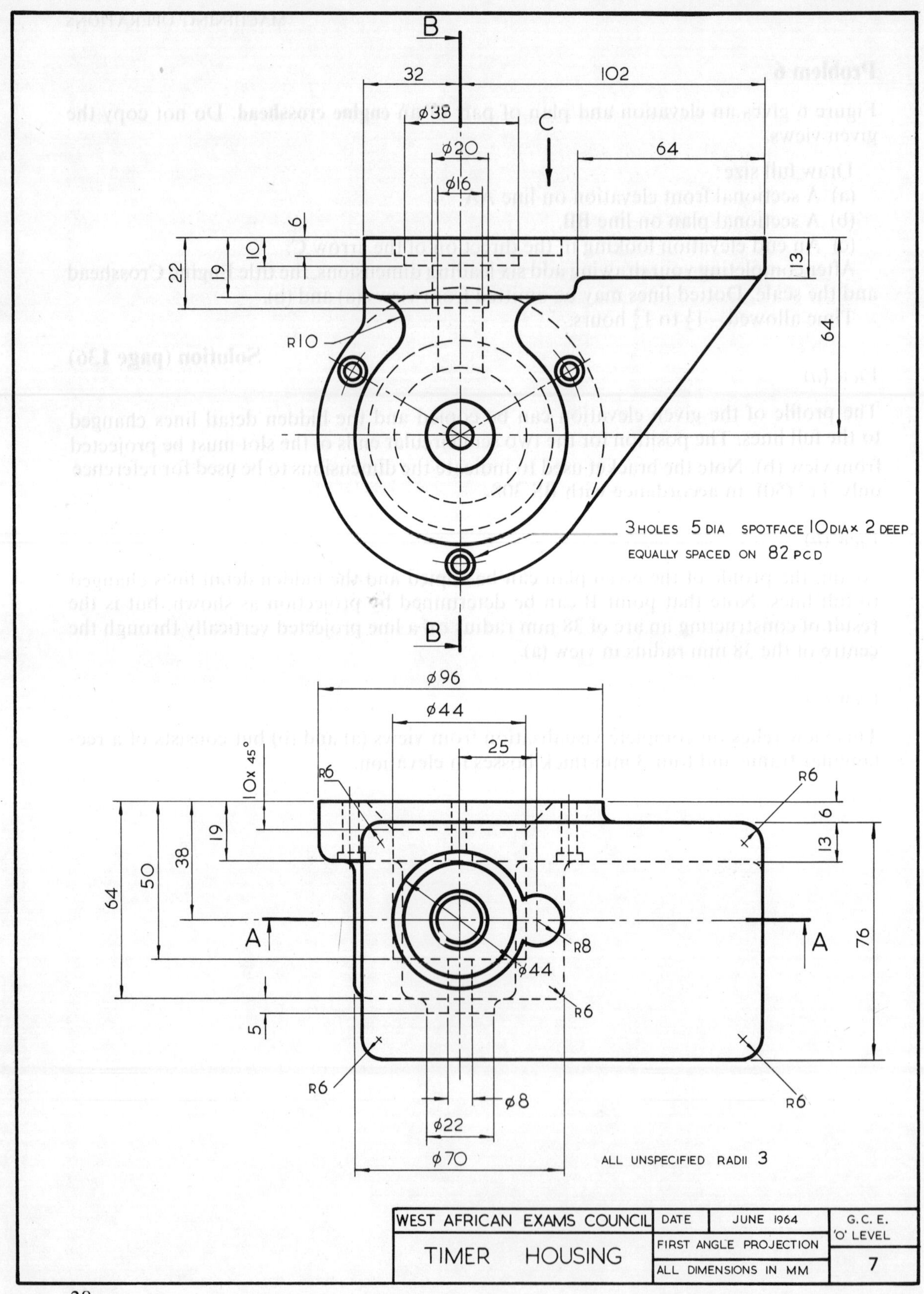

B
32
102
ø38
C
ø20
64
ø16
6
10
22
19
13
R10
64
3 HOLES 5 DIA SPOTFACE 10 DIA × 2 DEEP
EQUALLY SPACED ON 82 PCD
B
ø96
ø44
25
10 × 45°
R6
R6
6
19
13
38
50
64
A
A
76
R8
ø44
R6
5
R6
R6
ø8
ø22
ø70
ALL UNSPECIFIED RADII 3
WEST AFRICAN EXAMS COUNCIL
DATE
JUNE 1964
G.C.E.
'O' LEVEL
TIMER HOUSING
FIRST ANGLE PROJECTION
ALL DIMENSIONS IN MM
7

Problem 7

Views are given of a **timer housing** (Figure 7).

Draw, full size and in first angle projection:
(a) A sectional elevation, the plane of the section and the direction of the required view being indicated at AA.
(b) A sectional elevation, the plane of the section and the direction of the required view being indicated at BB.
(c) A plan, projected below view (b), looking in the direction of the arrow C.
Insert five important dimensions.
Hidden part lines are not required on any of the views.
Time allowed – $1\frac{1}{2}$ hours.

Solution (page 137)

View (a)

The pictorial view illustrates the component when sectioned by the cutting plane AA.
The majority of the component remains visible after being cut by the plane AA and as a result the profile of the given elevation can be copied. Careful study of the hidden detail shown in the given plan view is necessary to establish the bore of the circular part of the body cut by plane AA.

View (b)

The successful completion of view (a) should enable a straightforward projection to the right for view (b).
Note that the junction of the 16 mm diameter hole with the 44 mm diameter bore is as for intersecting cylinders.

View (c)

This view is a direct copy of the given plan view but is orientated differently to conform with the projection requirements of the question.

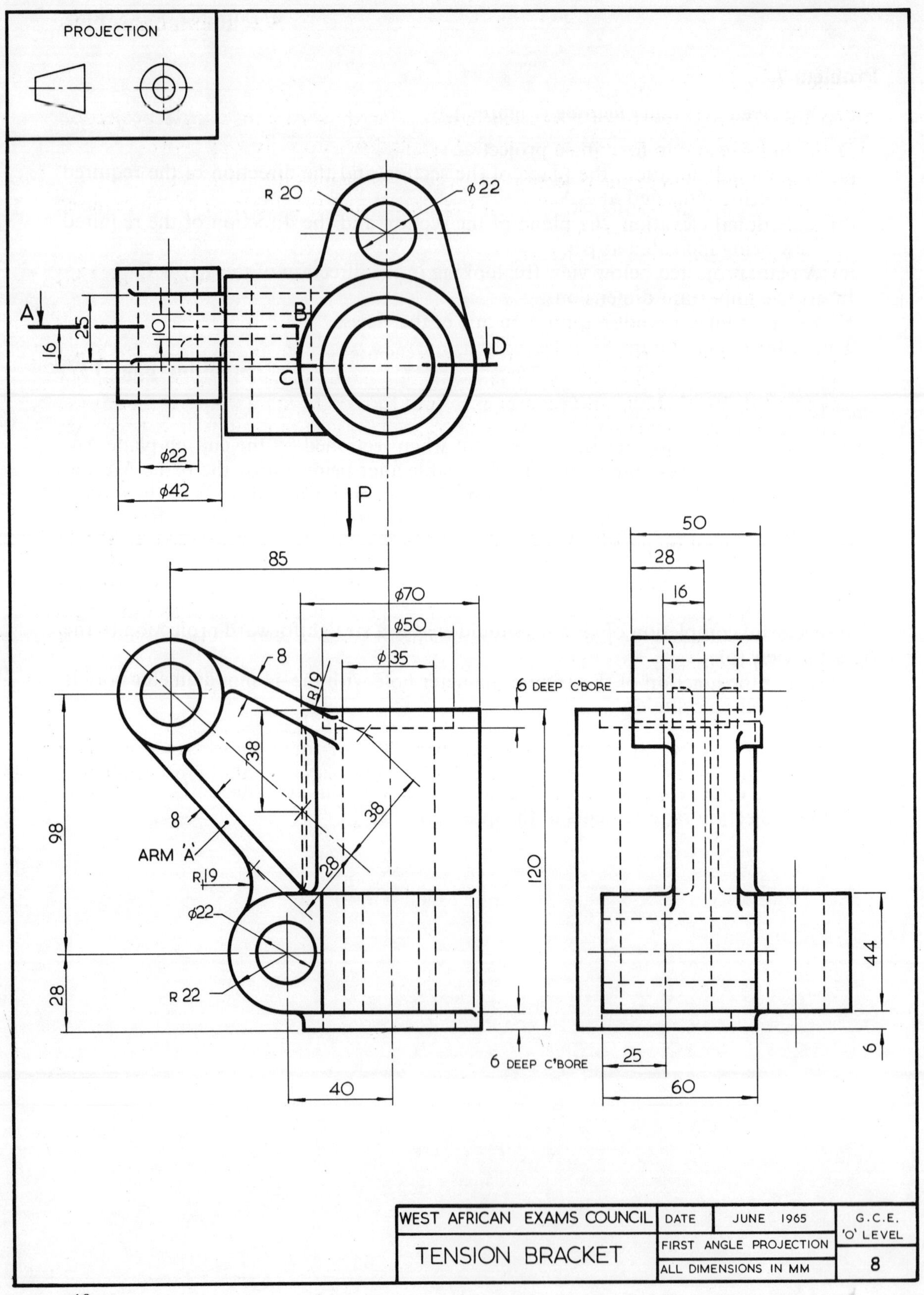
PROJECTION
R 20
ϕ22
A
B
C
D
25
10
16
ϕ22
ϕ42
P
85
ϕ70
ϕ50
ϕ35
8
R19
38
38
8
ARM 'A'
R 19
28
ϕ22
R 22
98
28
120
6 DEEP C'BORE
6 DEEP C'BORE
40
50
28
16
44
6
25
60
WEST AFRICAN EXAMS COUNCIL
DATE
JUNE 1965
G.C.E.
'O' LEVEL
TENSION BRACKET
FIRST ANGLE PROJECTION
ALL DIMENSIONS IN MM
8

Problem 8

Three orthographic views of a **tension bracket** are shown in first angle projection (Figure 8). Draw, full size, and in third angle projection, the following views:

(a) An outside plan in the direction of arrow P, showing all hidden detail.
(b) A sectional elevation, the plane of the section and the direction of the required view being indicated at ABCD.
(c) A simple web section of arm A.

Solution (page 138)

View (a)

Arrow 'P' indicates a viewing direction which is the opposite to the top left hand given view which is an inverted plan view. The profile of the required view will be similar but of the opposite hand about a horizontal line. Both ends of the main cylindrical body are machined in a similar manner but the Arm 'A' will be seen more fully in the view on arrow 'P'.

View (b)

For third angle projection this must be drawn beneath view (a). The section line ABCD is that of an offset section (see page 19) and the viewing direction is similar to that of the given front elevation (lower left hand). Note that the centre 10 mm thick web has not been sectioned in accordance with the requirements of BS 308, and that an angle of 30° has been chosen for the section lining. This is because Arm 'A' is inclined at an angle very close to 45°.

View (c)

This takes the form of a removed section (see page 20) and can be chosen at any position along the arm which includes the flanges. The section plane should be perpendicular to the centre line shown on the given front elevation.

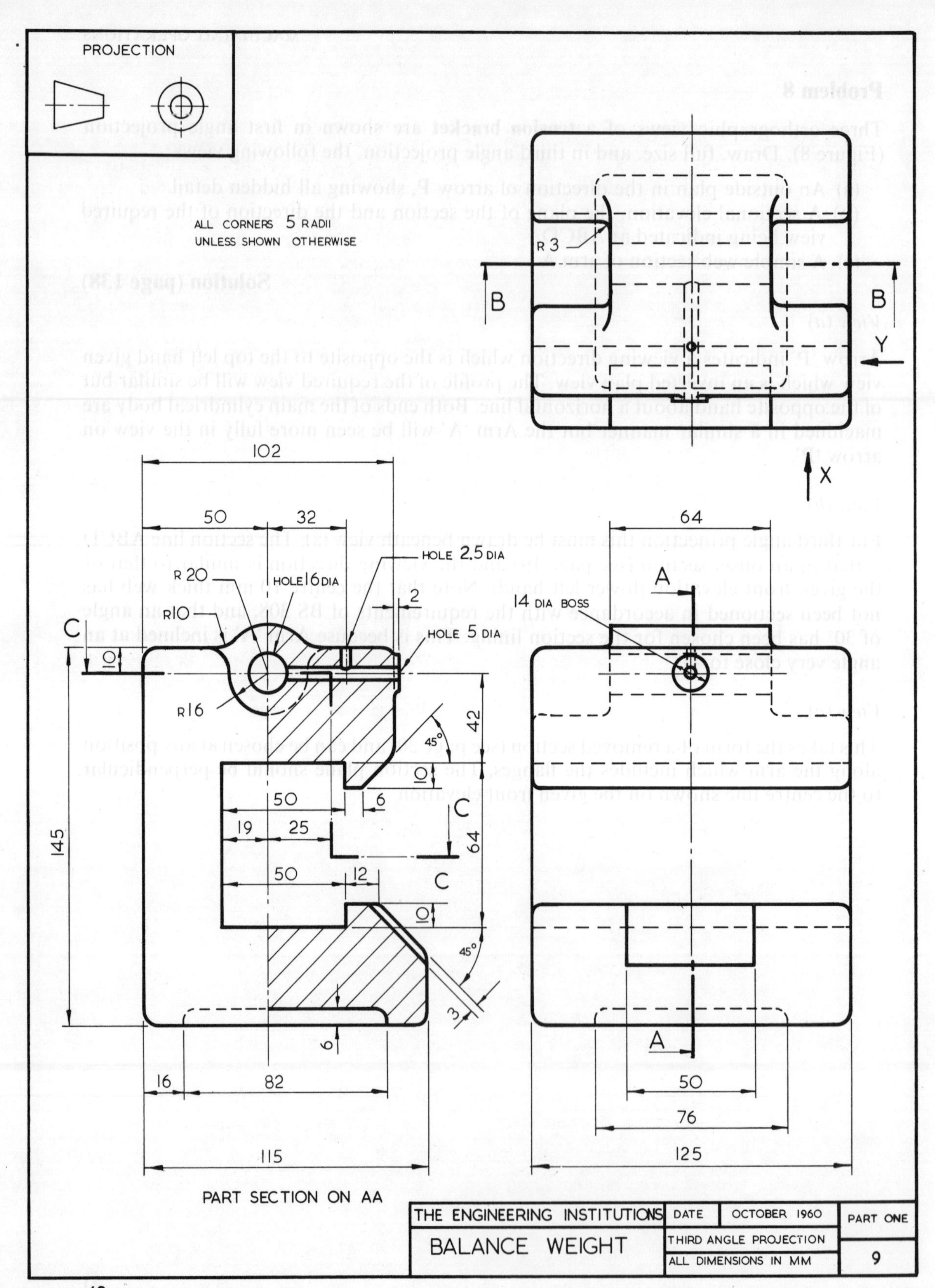

PROJECTION
ALL CORNERS 5 RADII
UNLESS SHOWN OTHERWISE
R 3
B
B
Y
X
102
50
32
64
HOLE 2.5 DIA
R 20
HOLE 16 DIA
2
R10
14 DIA BOSS
A
C
HOLE 5 DIA
10
R16
42
45°
10
50
6
C
19
25
145
64
50
12
C
10
45°
3
6
A
16
82
50
76
115
125
PART SECTION ON AA
THE ENGINEERING INSTITUTIONS
DATE
OCTOBER 1960
PART ONE
BALANCE WEIGHT
THIRD ANGLE PROJECTION
ALL DIMENSIONS IN MM
9

Problem 9

Figure 9 gives particulars of a **balance weight** for a weighing machine.

Draw full size the following views of the balance weight. Either first angle or third angle projection may be used, but the method of projection used must be stated.

(a) The half sectional elevation looking in the direction of the arrow X, showing the left hand half in outside view and the right hand half in section on BB.
(b) The outside elevation looking in the direction of the arrow Y.
(c) The sectional plan on CC.

Broken lines showing hidden detail are not required.

Time allowed – $1\frac{3}{4}$ hours.

Solution (page 139)

View (a)

The lower right hand given view is an outside elevation on arrow X and its left half is therefore similar to the left hand half of view (a). The half section on BB has a similar profile to the outside elevation on arrow X requiring only that the hidden detail lines be made full lines. The cutting plane BB should be identified in the given Part Section on AA as an aid to visualisation.

View (b)

The view on arrow Y is similar to the lower left hand given view titled Part Section on AA, but is of the opposite hand and depending on the angle of projection chosen is positioned to the left or right of view (a).

View (c)

The offset cutting plane CC results in a partial section only and ideally view (c) is positioned above or below view (b) depending on the angle of projection chosen. However section CC may be projected from view (a).

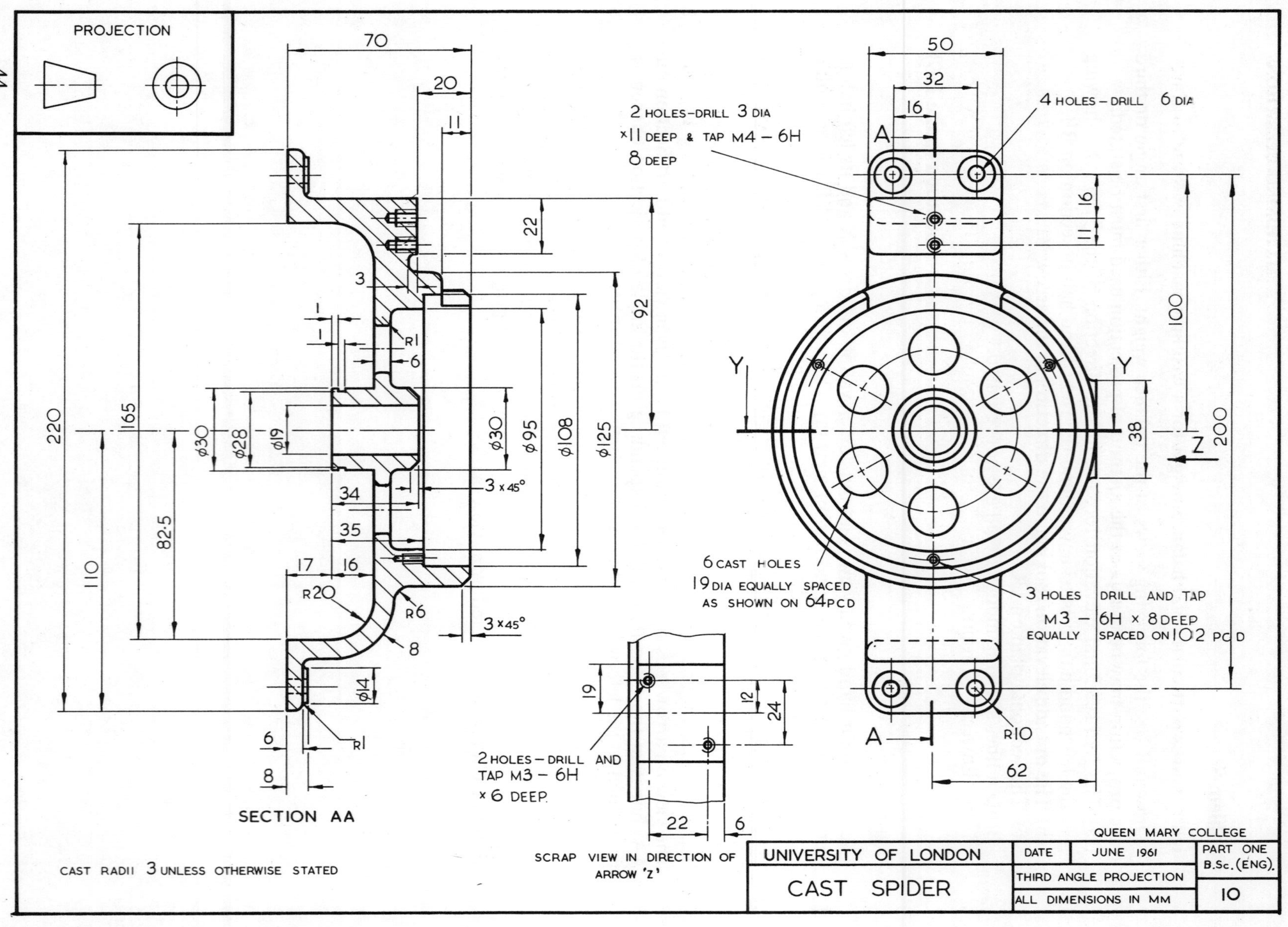

PROJECTION
SECTION AA
CAST RADII 3 UNLESS OTHERWISE STATED
2 HOLES-DRILL 3 DIA ×11 DEEP & TAP M4 – 6H 8 DEEP
4 HOLES – DRILL 6 DIA
6 CAST HOLES 19 DIA EQUALLY SPACED AS SHOWN ON 64 PCD
3 HOLES DRILL AND TAP M3 – 6H × 8 DEEP EQUALLY SPACED ON 102 PCD
2 HOLES – DRILL AND TAP M3 – 6H × 6 DEEP.
SCRAP VIEW IN DIRECTION OF ARROW 'Z'
QUEEN MARY COLLEGE
UNIVERSITY OF LONDON
CAST SPIDER
DATE
JUNE 1961
THIRD ANGLE PROJECTION
ALL DIMENSIONS IN MM
PART ONE B.Sc. (ENG).
10

Problem 10

Figure 10 shows two views and a scrap view of a **cast spider**. The views are in third angle projection.

The following views are required full size:

(a) An outside end elevation in direction of arrow Z.
(b) A half sectional plan as indicated by line YY.
(c) As much of the given front elevation as may be required for the construction of the two above views.

Dotted lines and dimensions need not be shown on the drawing.

Any dimensions which are necessary for the completion of the drawing but which are not given should be assessed by the candidate.

Either first angle or third angle projection may be used, but the method chosen should be clearly stated.

Time allowed – $1\frac{3}{4}$ hours.

Solution (page 140)

View (a)

This elevation is viewed in the opposite direction to that of the given Section AA and requires very little of the given front elevation (view (c)) for its completion. The only part necessary is that which shows the junction of the 50 mm wide leg with the 125 mm diameter body, and the profile of the body at the position of the 50 mm wide by 11 mm deep slot.

View (b)

It is necessary to include the 38 mm inch wide machined surface in view (c) in order that the position where this feature blends with the main 125 mm body can be projected into view (b).

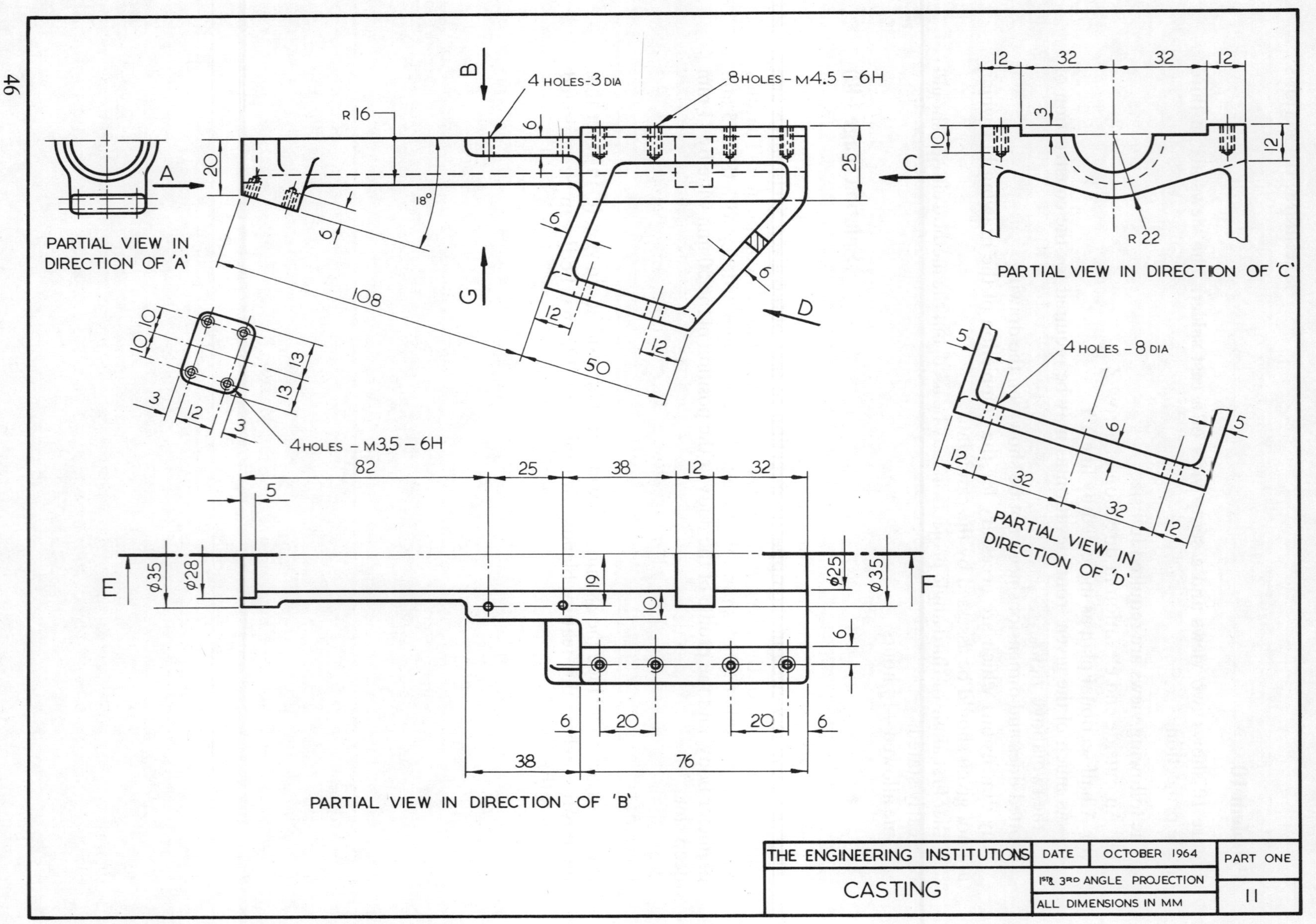
PARTIAL VIEW IN DIRECTION OF 'A'
A
B
G
C
D
E
F
R 16
4 HOLES-3 DIA
8 HOLES-M4.5 - 6H
20
6
18°
108
50
12
25
4 HOLES - M3.5 - 6H
10
13
3
PARTIAL VIEW IN DIRECTION OF 'C'
32
R 22
4 HOLES - 8 DIA
5
PARTIAL VIEW IN DIRECTION OF 'D'
82
38
ϕ35
ϕ28
ϕ25
19
76
PARTIAL VIEW IN DIRECTION OF 'B'
THE ENGINEERING INSTITUTIONS
DATE
OCTOBER 1964
PART ONE
CASTING
1ST & 3RD ANGLE PROJECTION
ALL DIMENSIONS IN MM
11

Problem 11

The details of a **casting** are as shown in Figure 11.

Draw, full size, the following views:

(a) A sectional elevation on EF.
(b) An outside view in direction G.
(c) An outside end elevation in direction C.

Insert five of the principal dimensions and add the title casting and the scale. First or third angle projection may be used, but the system must be stated.

Hidden detail need not be shown.

Time allowed – $1\frac{3}{4}$ hours.

Solution (page 141)

In the presentation of the question both first and third angle projection have been used. The two major views are in first angle whilst the partial views or auxiliary views have been drawn in third angle.

The use of partial views has enabled complete shape description of an unusual component with the minimum number of lines, and third angle projection enables the partial views to be more closely associated with the part which they are meant to describe.

View (a)

The section plane EF cuts through the longitudinal centre line of the component and the viewing direction results in view (a) having a similar profile to the given front elevation. The shape description is completed by substituting full lines for the hidden detail lines of the given elevation, with the exception of those for the tapped holes.

View (b)

Depending on the method of projection chosen view (b) will be positioned above or below view (a). Third angle has been chosen for the solution on page 141. This is an inverted plan view of the component presenting two faces not parallel to the viewing plane and which contain drilled and tapped holes. However, the angle is acute, only 18°, and the holes are best drawn with a compass.

Figure 11A shows the construction necessary for the line of intersection of the end pad.

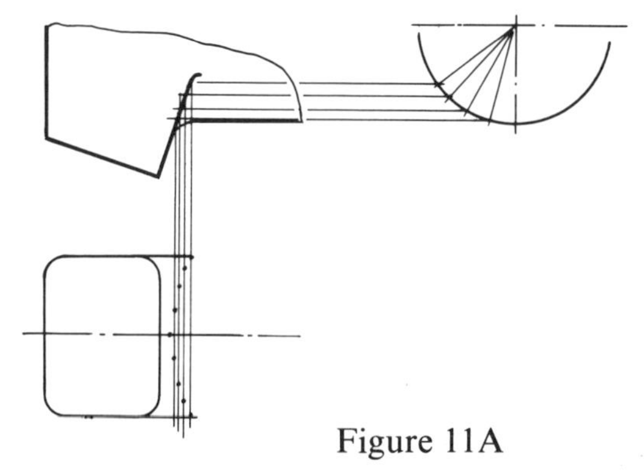

Figure 11A

View (c)

A part of this view has been given and requires only the projection from view (a) of the sloping bracket and end pad. The four 8 mm dia holes will appear elliptical in shape.

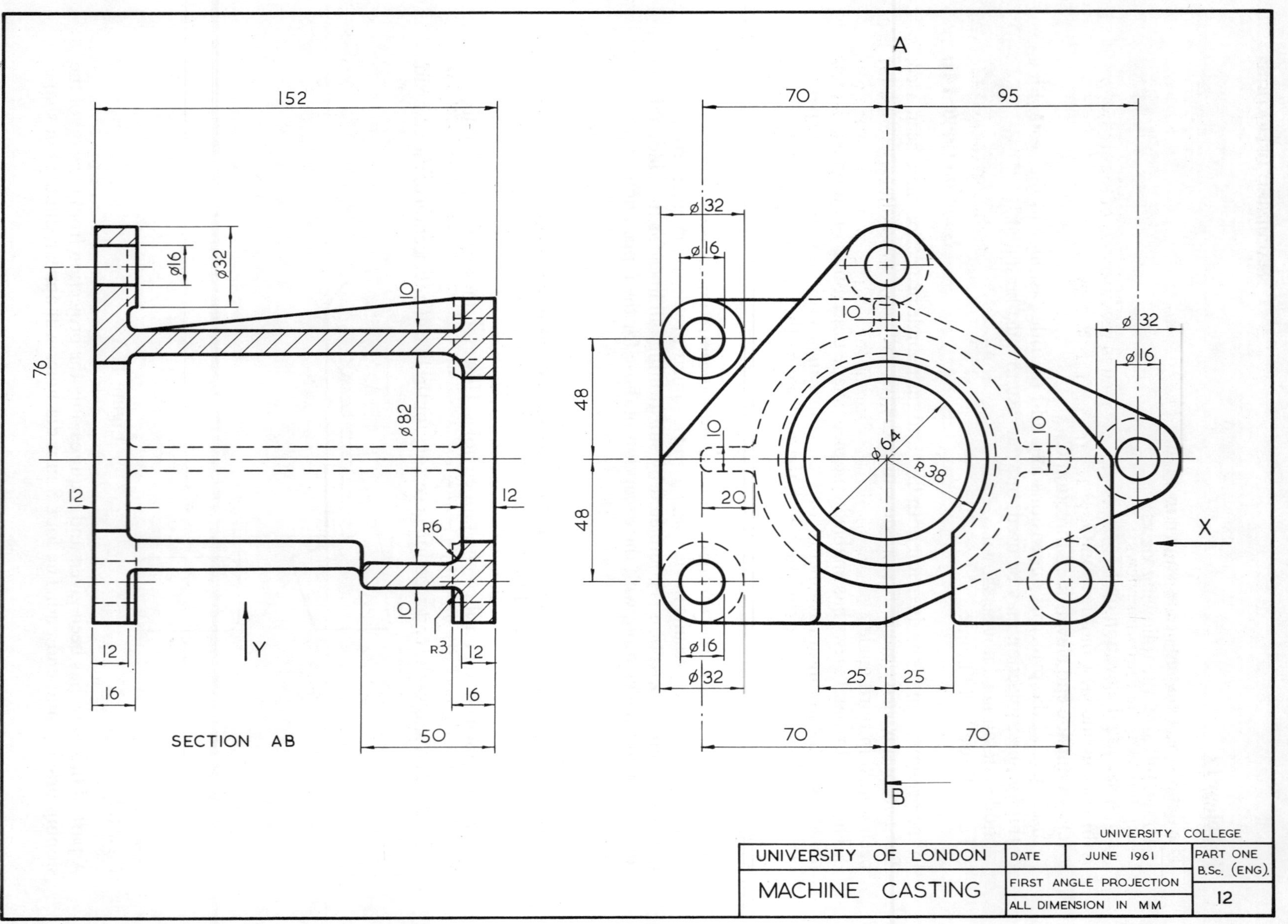
152
ϕ16
ϕ32
76
10
ϕ82
12
12
R6
10
R3
12
12
16
16
Y
SECTION AB
50
A
70
95
ϕ32
ϕ16
10
ϕ32
ϕ16
48
48
10
ϕ64
R38
10
20
X
ϕ16
ϕ32
25
25
70
70
B
UNIVERSITY COLLEGE
UNIVERSITY OF LONDON
DATE
JUNE 1961
PART ONE B.Sc. (ENG).
MACHINE CASTING
FIRST ANGLE PROJECTION
ALL DIMENSION IN MM
12

Problem 12

Figure 12 shows two views of a **machine casting**.

Draw, full size, the following views:
(a) In place of the section on AB, an outside view seen in the direction of the arrow X.
(b) A view in the direction of the arrow Y, projected from view (a).
Hidden edges are to be omitted from view (a) only.
Estimate any missing dimensions.
Time allowed – $1\frac{3}{4}$ hours.

Solution (page 142)

View (a)

Arrow X indicates a similar viewing direction to that of Section AB resulting in view (a) having a similar profile. As an aid to visualisation of the casting, particular attention should be given to the 32 mm diameter bosses; that is, with reference to which side of the flanges they are positioned.

In order that distance Z may be determined it is necessary to construct lightly a part of the right hand given elevation (Figure 12A).

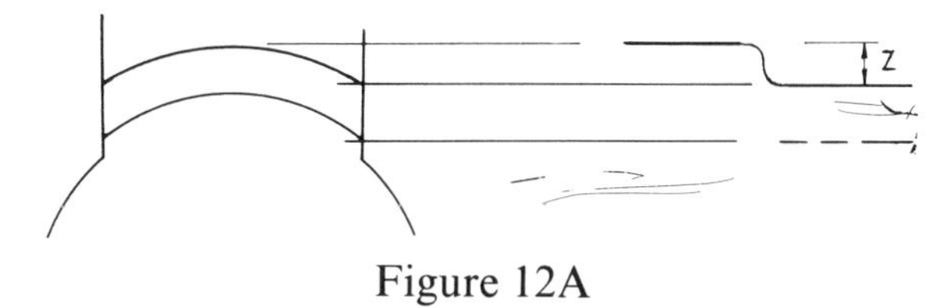

Figure 12A

Note that the top web in the given elevation has not been sectioned.

View (b)

As third angle projection has been chosen in the solution, view (b) is below view (a). The side webs are parallel since there is no hidden detail shown across their thickness in the given end elevation.

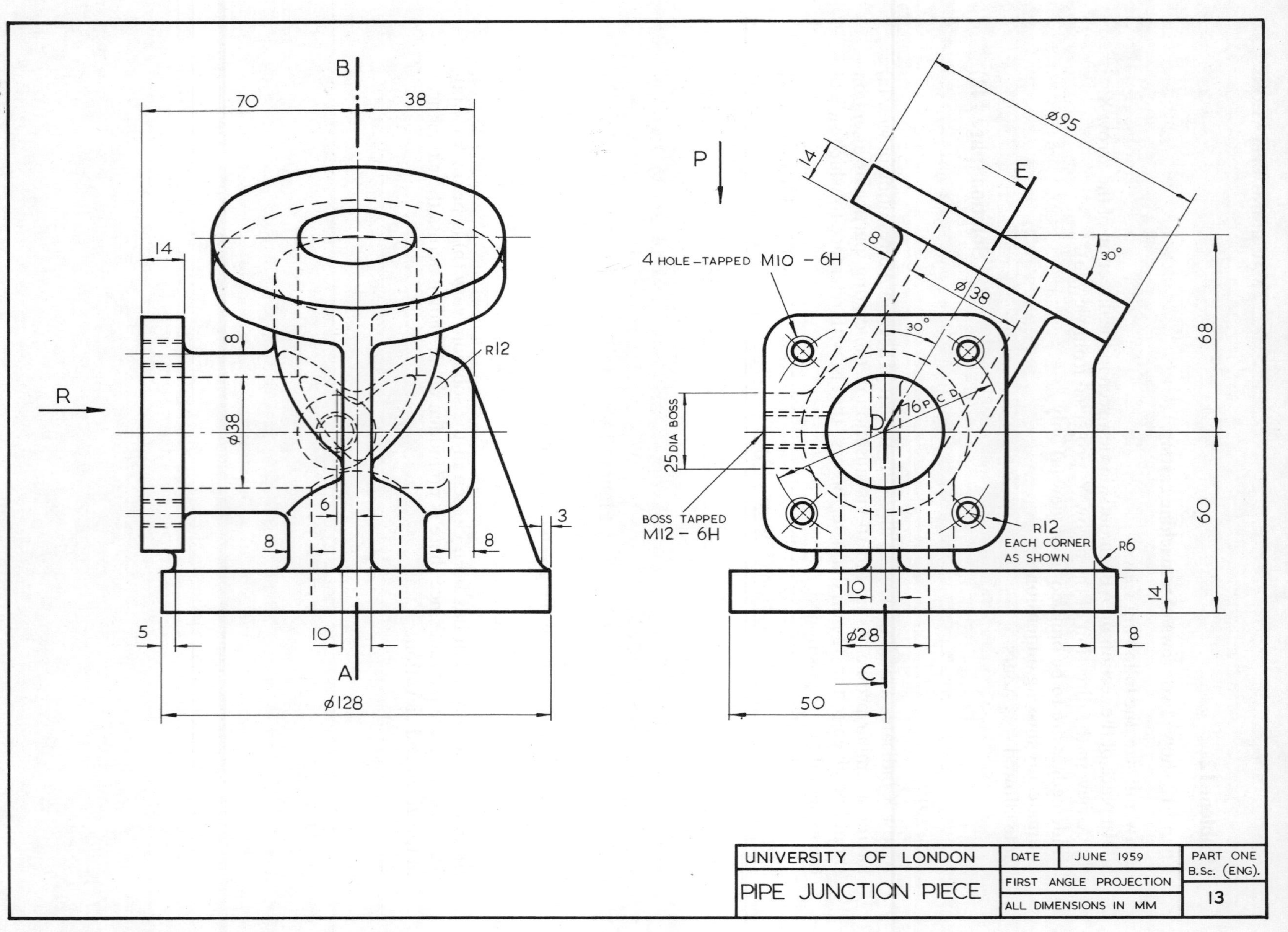

B
70
38
14
R12
R
ø38
8
6
3
5
10
A
ø128
P
14
E
ø95
8
4 HOLE-TAPPED M10 – 6H
30°
ø38
68
25 DIA BOSS
116 P C D
D
BOSS TAPPED
M12 – 6H
R12
EACH CORNER
AS SHOWN
R6
60
10
ø28
C
50
UNIVERSITY OF LONDON
PIPE JUNCTION PIECE
DATE
JUNE 1959
FIRST ANGLE PROJECTION
ALL DIMENSIONS IN MM
PART ONE
B.Sc. (ENG).
13

Problem 13

Two elevations of a **pipe junction piece** are shown in first angle projection in Figure 13.

Draw full size, the following views:

(a) A sectional elevation. The plane of the required section is indicated by the line AB and the view required is that seen when looking in the direction of arrow R.

(b) A sectional elevation drawn on the right of view (a), the planes of the required section being indicated by the lines CD, DE.

(c) A complete outside plan below view (a) and in projection with it. The view required is that seen when looking in the direction of arrow P.

Hidden edges are required on view (a) only. Insert on the drawing four important dimensions. Draw a title block 100 mm by 60 mm, in the lower right hand corner of the drawing paper, and insert the title and scale.

Time allowed – 2 hours.

Solution (page 143)

View (a)

Arrow R indicates a similar viewing direction to that of the given right hand elevation and hence Section AB has a similar profile. The exception being the square flange which is in front of the cutting plane. However, in order that the flange size may be determined and drawn in view (b) the centres of the M10 – 6H holes must be established.

It will be necessary to construct lightly, either as a scrap view or in view (b), the 25 mm diameter boss, in order that the size of the portion remaining behind the cutting plane can be projected into view (a).

A 10 mm thick web has been cut by the section plane but in a manner which requires that it is not cross-hatched. Hidden detail has been asked for in this view.

View (b)

View (b) is an example of an aligned section, i.e., the section as seen along DE is revolved through 30° to the vertical position and CDE is treated as a single plane. The section plane should be located in view (c) in order that the sectioned width of the 128 mm diameter base can be determined.

The two curves of intersection are as for perpendicular intersecting cylinders. The line of intersection of the bores of similar size can be determined by inspection, i.e., straight lines.

Two of the webs are cut by the section plane but in a manner which requires that they are not cross-hatched.

View (c)

The flange and bore diameters will appear as ellipses, and the two curves of intersection are of perpendicular intersecting cylinders.

6 SECTIONAL VIEWS (Assembly Drawings)

The introduction to Chapter 4 outlined the various methods of sectioning that can be used to clarify any internal feature of a component or generally help in conveying its complete shape description. Assembly drawings can be treated in a similar way in as much as sectional views can be used to give a clear picture of the function of each part and the manner in which is is assembled.

The recommendations made previously for single components apply to sectional views of assemblies but it is necessary to make a number of additional rules.

Adjacent parts

The section lining of adjacent parts should be in opposite directions (Figure 27A). Where this is not possible, as in the case of component A which must have section lining in a similar direction to either part B or C, then the pitch of the section lines are made different (Figure 27B).

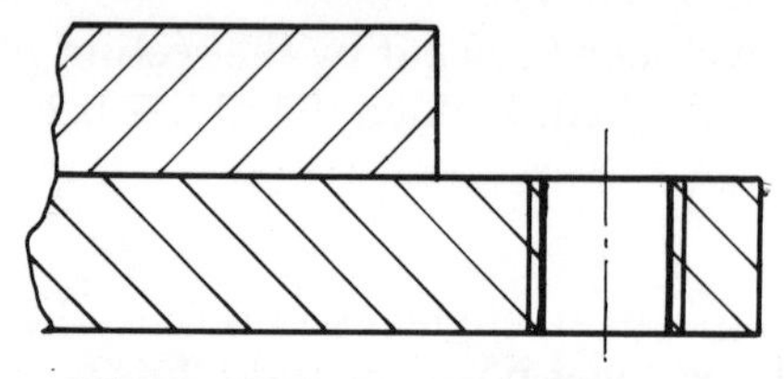

Figure 27A

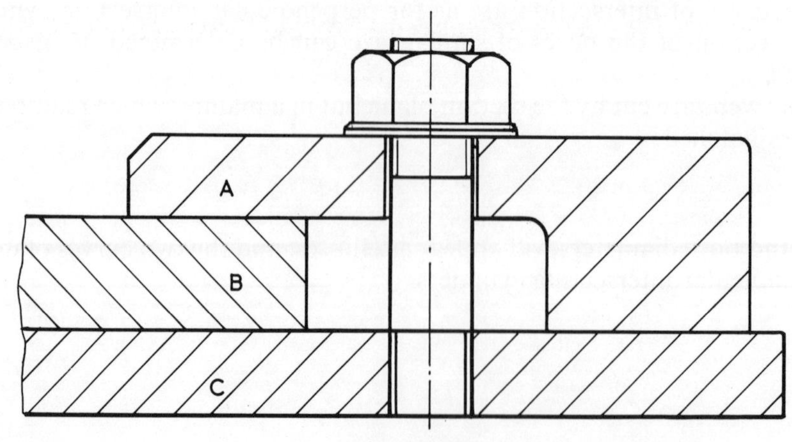

Figure 27B

Parts not sectioned

When a cutting plane passes through shafts, bolts, nuts, rods, rivets, keys, pins and similar standard parts, it is not normal to show them in section. Remember that the purpose of a sectional view is to clarify the arrangement of assembled parts, and the sectioning of parts similar to those mentioned would not normally assist in any way. Figure 27C illustrates the conventional way of sectioning such an assembly.

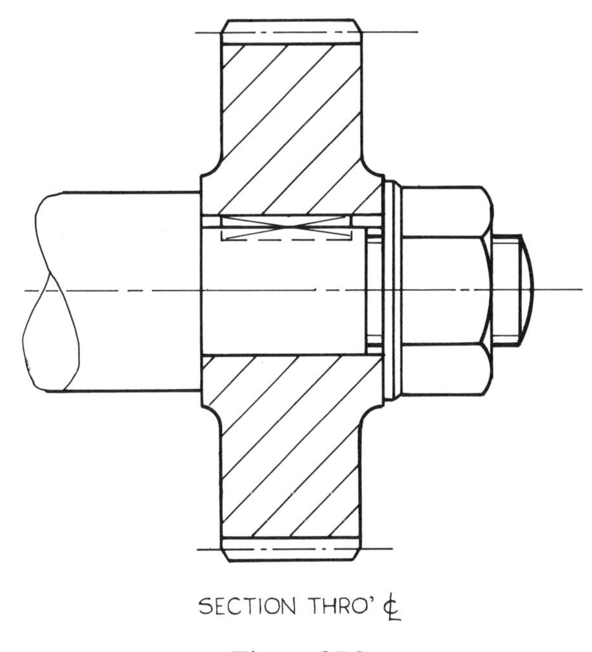

SECTION THRO' ℄

Figure 27C

PROBLEMS 14 to 20

This section contains assembly drawings of increasing difficulty.

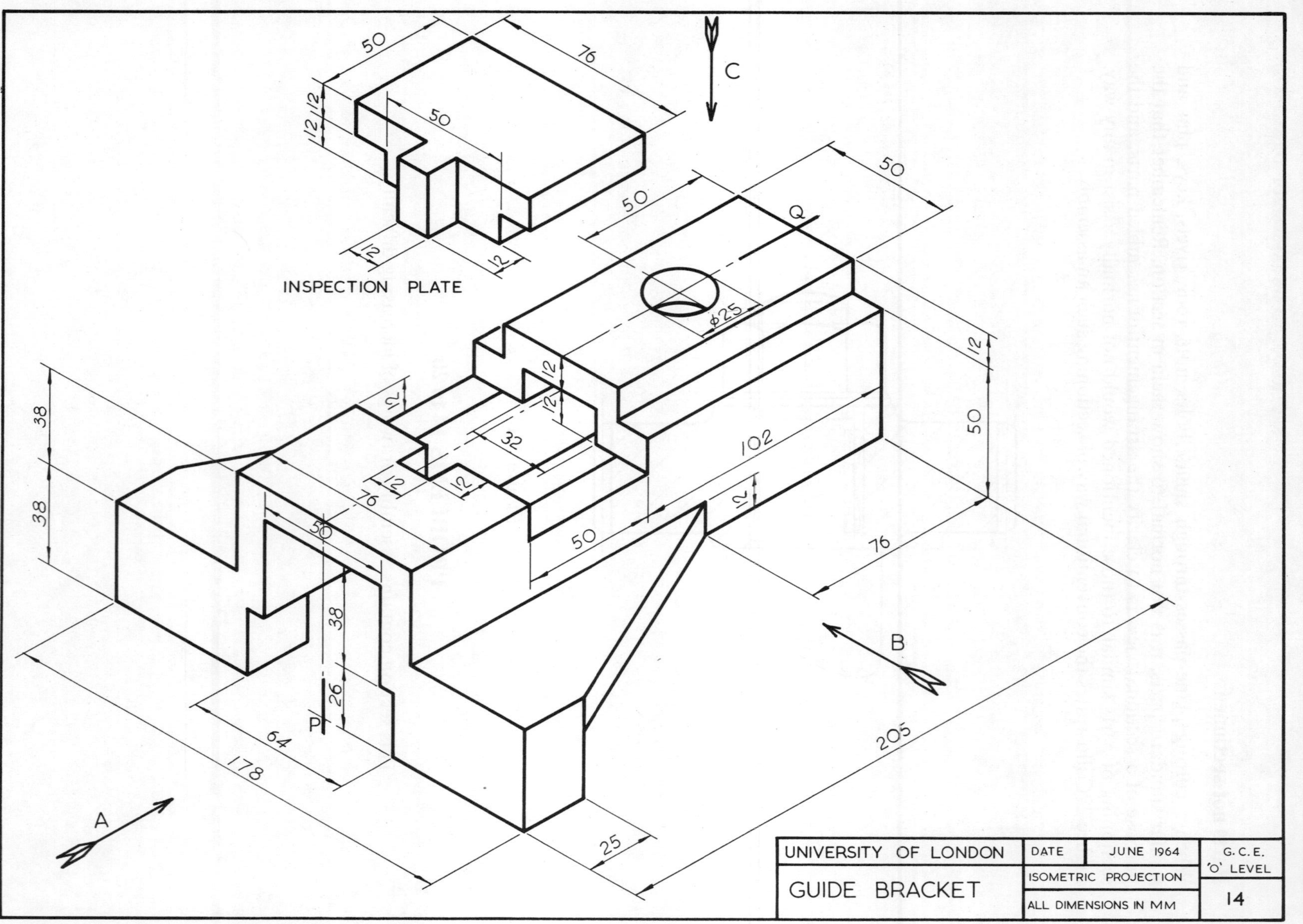
INSPECTION PLATE
C
A
B
P
Q
ϕ25
50
76
50
12
12
12
12
50
50
12
12
12
12
32
12
12
50
102
76
50
38
38
38
26
64
178
205
76
25
12
UNIVERSITY OF LONDON
GUIDE BRACKET
DATE
JUNE 1964
ISOMETRIC PROJECTION
ALL DIMENSIONS IN MM
G.C.E.
'O' LEVEL
14

Problem 14

Figure 14 shows a pictorial view of a **guide bracket** which incorporates a separate inspection plate. Draw full size and in correct orthographic projection the following views of the guide bracket with the inspection plate fitted into its slot:

(a) An elevation in direction of arrow A.
(b) A sectional elevation on plane PQ and in the direction of arrow B.
(c) A plan in the direction of arrow C.

Show in views (a) and (c) all hidden edges and include on your drawing six important dimensions.

Either first angle or third angle (but not both) methods of projection may be used; the method chosen must be stated on the drawing.

Time allowed – $1\frac{1}{2}$ hours.

Solution (page 144)

View (a)

Visualisation of the components has been helped by presenting them with pictorial views. Having identified the surfaces which come in contact, the two parts can be assembled. In view (a) the inspection plate is seen only inside the hollow section through the centre of the main bracket.

View (b)

This view illustrates how the separate parts have been identified by cross-hatching in opposite directions. Note that the question does not ask for hidden detail in this view. Only in special cases is hidden detail shown in sectional views.

View (c)

In this view – and indeed in views (a) and (b) – it can be seen that a single line is used to convey the shape description of separate surfaces in contact. In order that parts can be assembled a clearance must be allowed but this is not shown since it will be very small indeed.

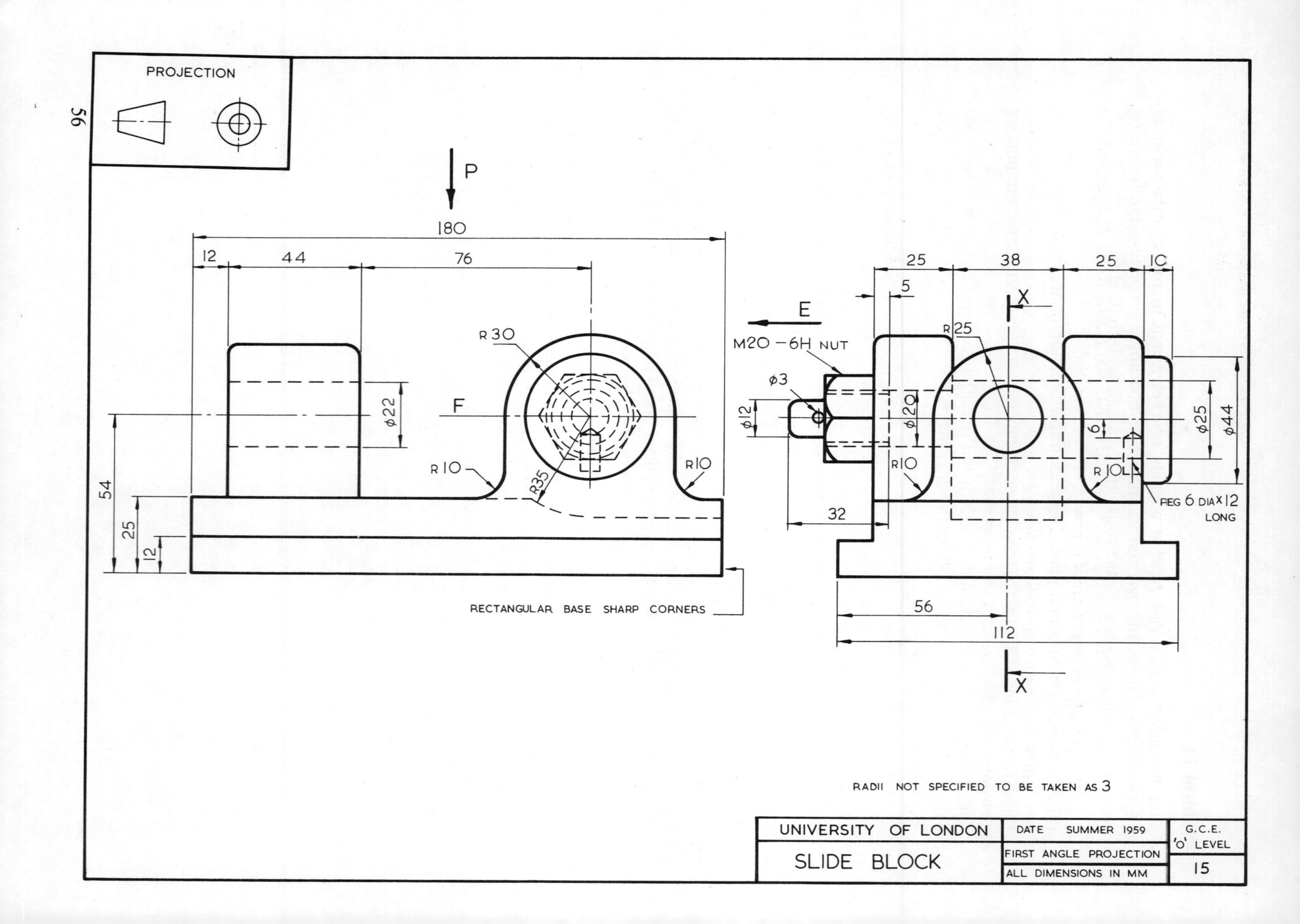

PROJECTION
P
180
12
44
76
R 30
F
ϕ22
R 10
R35
R10
54
25
12
RECTANGULAR BASE SHARP CORNERS
E
M20 - 6H NUT
ϕ3
ϕ12
25
5
38
25
IC
X
R25
ϕ20
6
ϕ25
ϕ44
R10
R 10
PEG 6 DIA X 12 LONG
32
56
112
X
RADII NOT SPECIFIED TO BE TAKEN AS 3
UNIVERSITY OF LONDON
SLIDE BLOCK
DATE SUMMER 1959
FIRST ANGLE PROJECTION
ALL DIMENSIONS IN MM
G.C.E. 'O' LEVEL
15

Problem 15

Orthographic views of a **slide block** are shown in Figure 15. Do not copy the views as shown but draw the following views:

(a) In place of the elevation F a sectional elevation, the plane of the section and the direction of the required view being indicated at XX.
(b) A complete outside end elevation in the direction of arrow E.
(c) A complete outside plan in the direction of arrow P and in projection with view (a).

Hidden part lines are not required on any of the views. Insert on the drawing six dimensions, the title, and the scale.

Time allowed – $1\frac{1}{2}$ to $1\frac{3}{4}$ hours.

Solution (page 145)

View (a)

Although a shaft is one of the parts of an assembly not normally sectioned, when the section plane cuts across its section as in view (a), then it is quite correct to show it cross-hatched. Indeed if it were not it would not be apparent that a shaft existed at all.

View (b)

Arrow E indicates the reverse viewing direction to that of the given end elevation. The across flats view of the nut will necessitate the construction of a hexagon and the use, in an examination, of the application of the relationship – across corners = 2D where D is the diameter of the bolt. The nut thickness can be made equal to 0·8D.

The hexagon should be drawn lightly adjacent to its elevation in view (b).

View (c)

The across corners dimension of the nut can be determined from the details provided in view (b). The lines of intersection extending from the lug at the left hand end are as a result of the 10 mm radii.

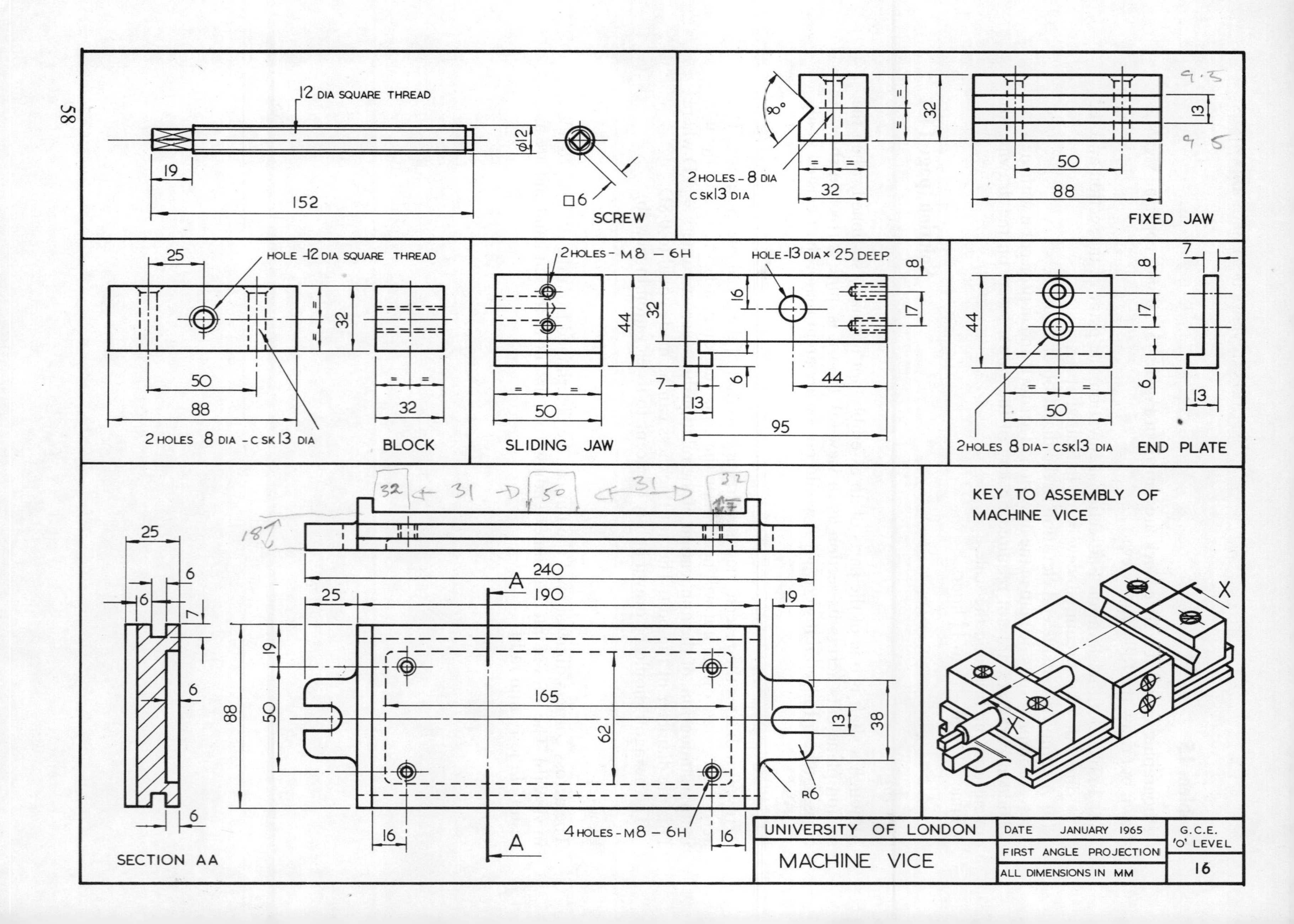
12 DIA SQUARE THREAD
SCREW
FIXED JAW
2 HOLES - 8 DIA C SK 13 DIA
HOLE - 12 DIA SQUARE THREAD
2 HOLES 8 DIA - C SK 13 DIA
BLOCK
2 HOLES - M8 - 6H
HOLE - 13 DIA × 25 DEEP
SLIDING JAW
2 HOLES 8 DIA - CSK 13 DIA
END PLATE
KEY TO ASSEMBLY OF MACHINE VICE
4 HOLES - M8 - 6H
SECTION AA
UNIVERSITY OF LONDON
MACHINE VICE
DATE JANUARY 1965
FIRST ANGLE PROJECTION
ALL DIMENSIONS IN MM
G.C.E. 'O' LEVEL
16

Problem 16

Figure 16 shows the details of a small **machine vice** and a key to its assembly.

Draw, full size, and in correct orthographic projection the following views of the completely assembled vice, the sliding jaw being approximately 25 mm from the fixed jaw.

(a) A sectional elevation on a vertical plane passing through the axis of the square threaded screw, in the direction indicated by XX in the key.

(b) A plan projected from the above.

Details of screws are not given, and these may be omitted from your drawing. No hidden edges are to be shown and dimensions are not required.

Either first angle or third angle (but not both) methods of projection may be used; the method chosen must be stated on the drawing.

Time allowed – 2 hours.

Solution (page 146)

View (a)

The pictorial view shows clearly how each part is assembled and all features of the base are clearly shown in the given views. The recess on the underside of the base casting has been dimensioned at the hidden detail of the plan view, i.e., 165 mm × 62 mm. The position of the block and fixed jaw are controlled by their clearance holes and the tapped holes in the base. In accordance with recommendations of BS 308 the screw has not been sectioned.

View (b)

Third angle projection has been chosen in the solution and requires that view (b) is positioned above view (a). The assembly of the sliding jaw provides for the end plate to be assembled on the nearest side on view (b).

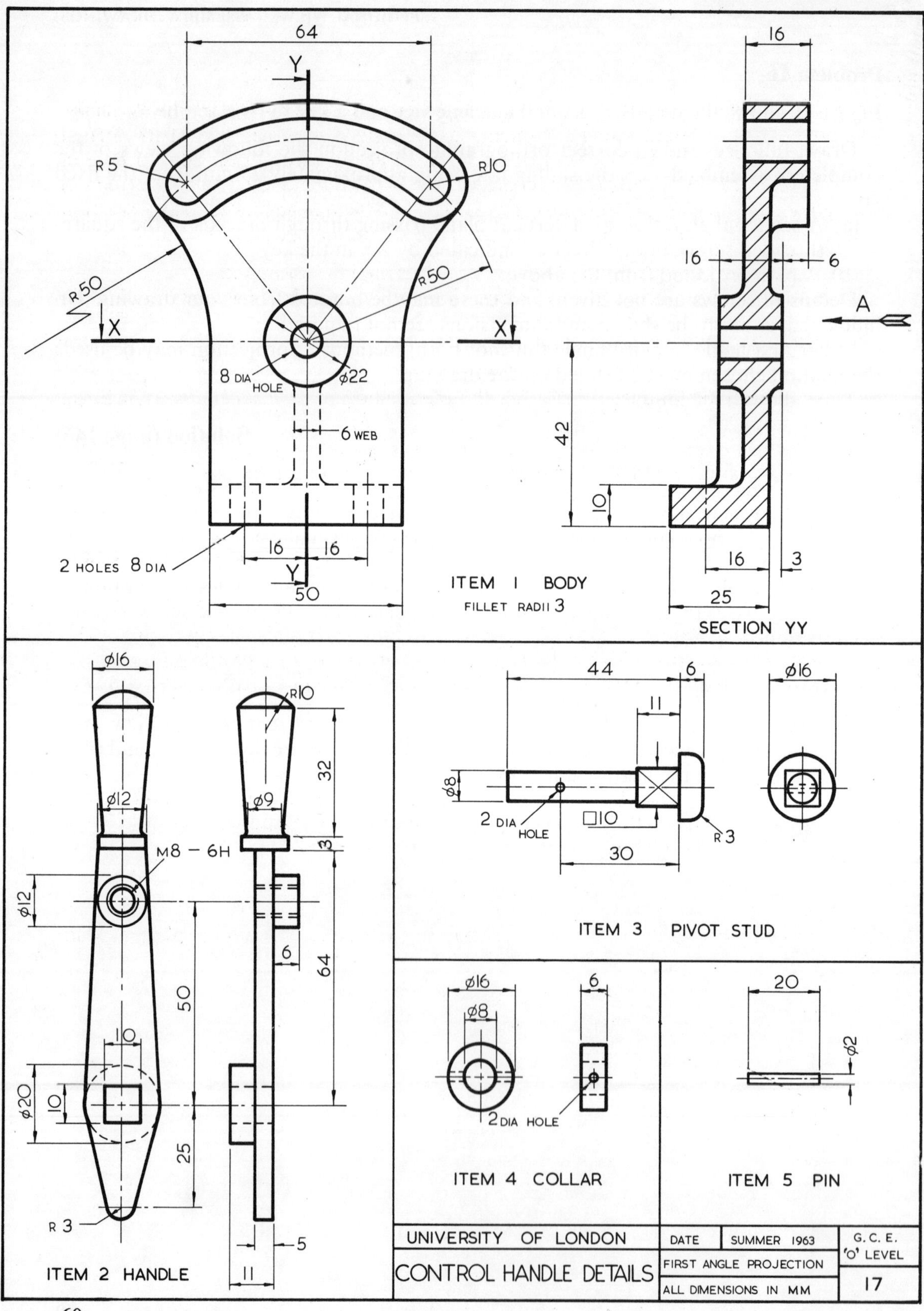
64
16
Y
R5
R10
16
6
A
R50
R50
X
X
8 DIA HOLE
⌀22
6 WEB
42
10
2 HOLES 8 DIA
16
16
Y
50
16
3
25
ITEM 1 BODY
FILLET RADII 3
SECTION YY
⌀16
R10
32
⌀12
⌀9
M8 – 6H
3
⌀12
6
64
50
10
⌀20
10
25
R 3
5
11
ITEM 2 HANDLE
44
6
11
⌀8
2 DIA HOLE
□10
R 3
30
⌀16
ITEM 3 PIVOT STUD
⌀16
⌀8
6
2 DIA HOLE
ITEM 4 COLLAR
20
⌀2
ITEM 5 PIN
UNIVERSITY OF LONDON
DATE
SUMMER 1963
G. C. E. 'O' LEVEL
CONTROL HANDLE DETAILS
FIRST ANGLE PROJECTION
ALL DIMENSIONS IN MM
17

Problem 17

Figure 17 shows the five detail parts which together make up a **control handle assembly**. This consists of a handle (item 2) which may be moved to either side of its normal vertical position about a pivot stud (item 3) which passes through the central hole in the body (item 1). The pivot stud is secured at the back of the body by a collar (item 4) through which is passed a 2 mm pin (item 5). The handle may be locked in any position by a screw (not shown) which passes through the slot in the body.

Any dimensions not given on the detail drawings may be estimated.

Draw in correct projection the following views of the completely assembled unit, the handle being in the vertical position:

(a) Front elevation in direction of arrow A.
(b) Elevation projected to the left of (a).
(c) Sectional plan projected from (a), the plane and direction of the section being shown at XX.

No dimensions are to be shown, and hidden edges are not required. Either first angle or third angle method of projection (but not both) may be used; the method chosen is to be shown clearly in the drawing.

Time allowed – $1\frac{3}{4}$ hours.

Solution (page 147)

View (a)

The given front elevation of the body is a view on arrow A and consequently can be reproduced for view (a). The handle is assembled on this front face and similarly the given front elevation of the handle can be reproduced in the vertical position plus the circular head of the pivot stud.

View (b)

Because the question demands that view (b) is projected to the left of view (a) the 'hand' of view (b) will depend on the method of projection chosen. The 'run out' position of the main vertical plate of the body can be projected from view (a).

View (c)

This section illustrates the convention relating to shafts, etc., by leaving the pivot stud and pin unsectioned.

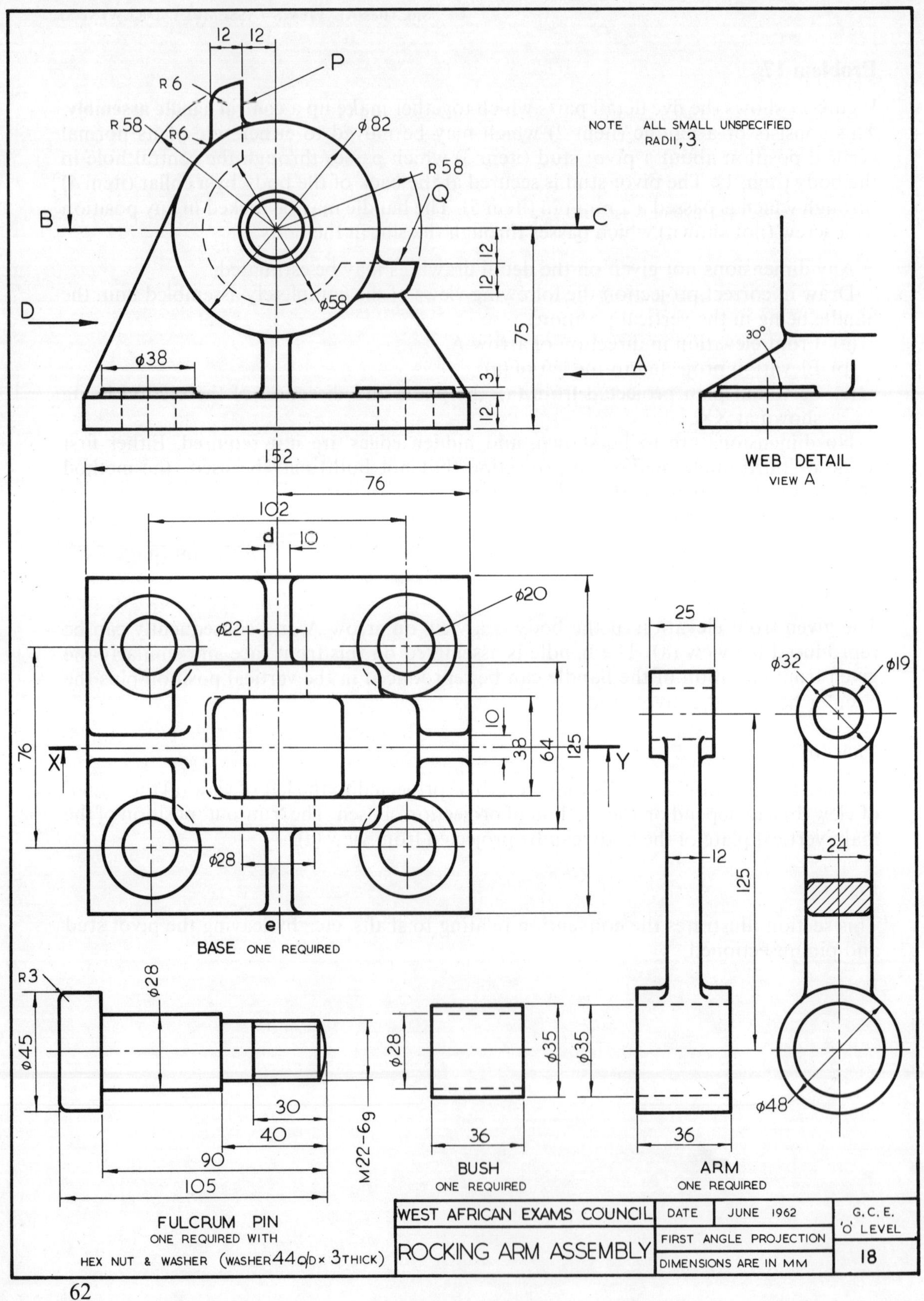
NOTE ALL SMALL UNSPECIFIED RADII, 3.
P
R6
R58
R6
ϕ82
R58
Q
ϕ58
B
C
D
A
ϕ38
75
12
12
3
12
152
76
102
d
10
WEB DETAIL
VIEW A
30°
ϕ20
ϕ22
76
X
Y
10
38
64
125
ϕ28
e
BASE ONE REQUIRED
25
ϕ32
ϕ19
12
125
24
R3
ϕ28
ϕ45
30
40
90
105
M22-6g
ϕ28
ϕ35
ϕ35
36
36
ϕ48
FULCRUM PIN
ONE REQUIRED WITH
HEX NUT & WASHER (WASHER 44 o/d × 3 THICK)
BUSH
ONE REQUIRED
ARM
ONE REQUIRED
WEST AFRICAN EXAMS COUNCIL
DATE
JUNE 1962
G.C.E. 'O' LEVEL
FIRST ANGLE PROJECTION
ROCKING ARM ASSEMBLY
DIMENSIONS ARE IN MM
18

Problem 18

Details are given in Figure 18 of a **rocking arm assembly**.

Particulars of the Assembly

The bush is pressed into the larger end of the arm. These are then mounted on the fulcrum pin, on the centre line 'de' of the base, and secured by means of a 22 mm standard hexagonal nut. The arm should then be free to oscillate through 90° between the stops P and Q of the base.

Draw, full size:

(a) An elevation of the rocking arm assembly, with the arm resting against the stop Q. The elevation is to be in section through XY.
(b) A plan of the rocking arm assembly. The plan is to be in section through BC.
(c) An end elevation of the rocking arm assembly. The end elevation is to be a view looking in the direction of the arrow D.

Broken lines representing hidden details are not required in any of the views.
Insert six leading dimensions and the scale, and letter the title 'Rocking Arm Assembly'
Time allowed – $1\frac{3}{4}$ hours.

Solution (page 148)

View (a)

Here the centre supporting web which is parallel to the cutting plane has not been sectioned in accordance with BS 308. The connecting rib of the arm (24 mm × 12 mm) has been sectioned because the 12 mm thickness is a main member of the component.

View (b)

The arm has a 2 mm clearance when assembled in the base and this should be equally divided on each side. To simplify the construction of the full nut in views (b) and (c) a hexagon should be constructed lightly in view (a).

View (c)

The 'run out' position of the centre web can be projected from view (a).

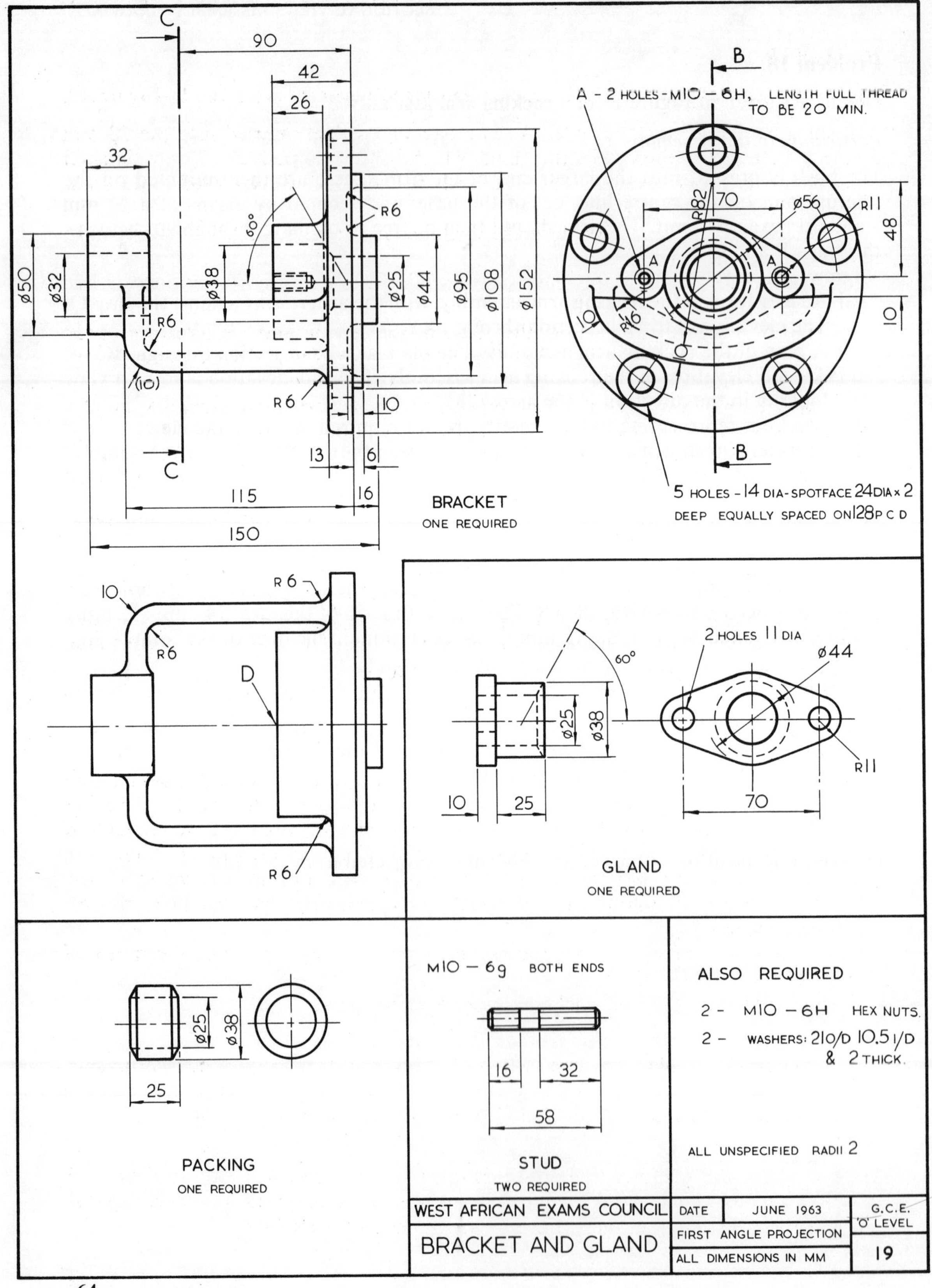
A - 2 HOLES - M10 - 6H, LENGTH FULL THREAD TO BE 20 MIN.
5 HOLES - 14 DIA - SPOTFACE 24DIA x 2 DEEP EQUALLY SPACED ON 128 P C D
BRACKET
ONE REQUIRED
2 HOLES 11 DIA
GLAND
ONE REQUIRED
PACKING
ONE REQUIRED
M10 - 6g BOTH ENDS
STUD
TWO REQUIRED
ALSO REQUIRED
2 - M10 - 6H HEX NUTS.
2 - WASHERS: 21 O/D 10.5 I/D & 2 THICK.
ALL UNSPECIFIED RADII 2
WEST AFRICAN EXAMS COUNCIL
DATE
JUNE 1963
G.C.E. 'O' LEVEL
FIRST ANGLE PROJECTION
BRACKET AND GLAND
ALL DIMENSIONS IN MM
19

Problem 19

Detail drawings of the parts of a **bracket and gland assembly** are given in Figure 19.

The studs are fitted into the holes A. The packing is then inserted into the 38 mm diameter hole at D, followed by the gland, which is fitted so that the 11 mm diameter holes pass over the projecting studs. The washers and nuts are then fitted to the studs and the nuts tightened.

Do not draw the given views but make the following views, full size, of the assembly, in first angle projection:

(a) A sectional elevation, the plane of the section and the direction of the required view being indicated at BB.

(b) Half of a sectional elevation, the plane of the section and the direction of the required view being indicated at CC. Draw that half which appears on the left of the vertical centre line.

(c) A half plan. Draw that half which appears above the horizontal centre line, in plan.

Broken lines representing hidden details are not required in any of the views.

Time allowed – $1\frac{3}{4}$ hours.

Solution (page 149)

When assembled, the bracket, gland, packing and securing studs and nuts form a unit similar to that of a stuffing box.

The packing, usually made from asbestos cord, when compressed by tightening the nut provides an efficient seal with the 25 mm diameter valve spindle (not shown).

The stud end with a thread length of 16 mm is assembled in the bracket at the holes tapped with a full thread to a minimum depth of 20 mm. This will ensure correct stud assembly, i.e., the stud cannot 'bottom' in the tapped hole.

The three views required result in the assembly being viewed in similar directions to the brackets three given views.

View (a)

Three parts become adjacent in section, i.e., bracket, gland and packing, demanding that the cross-hatch lines be varied in direction of slope and in their spacing. BS 308 recommends that glass, wood, insulation, concrete and water are illustrated in section by their own conventions and that ordinary section lining be used in all other cases where materials are shown in section. The method used to illustrate the sectioned packing in the model solution is one which some industries have adopted, not an accepted standard.

The position of the end of the stud can be determined by measuring a distance of 42 mm from the face of the bracket.

It can be assumed that full thickness nuts are to be used.

View (b)

The hexagon nut should be drawn in this view first and then be projected into views (a) and (c).

Remember to show the stud in plan view and not the thread of the nut.

View (c)

The position of the gland can be projected from view (a), together with details of the stud and nut. Remember that the corners are seen to be removed from a hexagonal nut when viewed across corners, but not when viewed across flats.

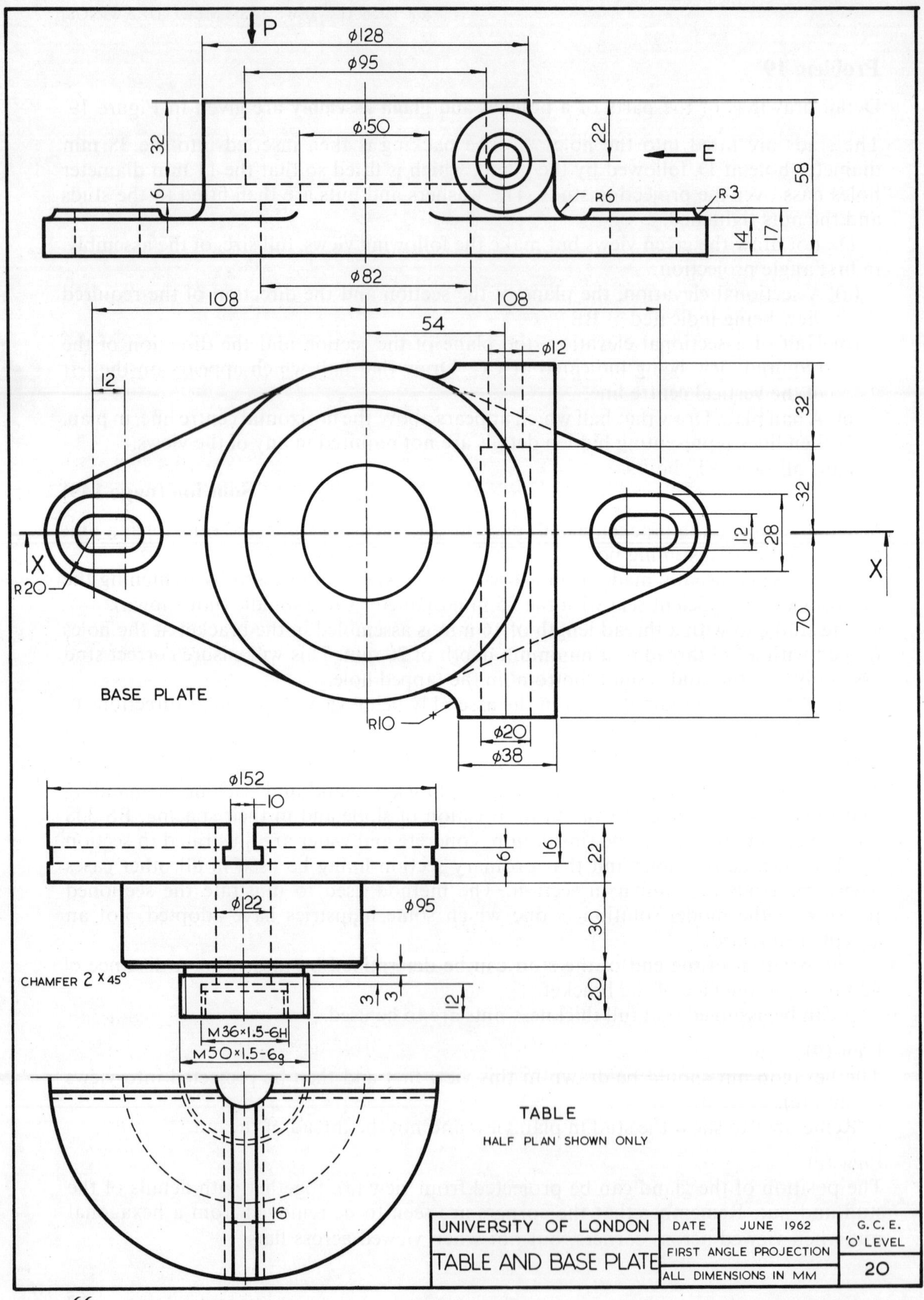
P
E
X
X
ϕ128
ϕ95
ϕ50
22
32
58
6
R6
R3
14
17
ϕ82
108
108
54
ϕ12
12
32
32
12
28
70
R20
BASE PLATE
R10
ϕ20
ϕ38
ϕ152
10
16
22
6
ϕ22
ϕ95
30
20
3
3
12
CHAMFER 2 × 45°
M36×1.5-6H
M50×1.5-6g
TABLE
HALF PLAN SHOWN ONLY
16
UNIVERSITY OF LONDON
TABLE AND BASE PLATE
DATE
JUNE 1962
FIRST ANGLE PROJECTION
ALL DIMENSIONS IN MM
G.C.E.
'O' LEVEL
20

Problem 20

Details of a circular **table** and its **base plate** are in Figure 20.

Do not copy the views as shown but draw, full size, the following views with the table in its working position on the base plate:

(a) A sectional elevation, the plane of the section and the direction of the required view being indicated at XX.
(b) An elevation in projection with view (a) looking in the direction of the arrow E.
(c) A complete plan projected from view (a) and looking in the direction of the arrow P.

No hidden edges need be shown.

Insert on the drawing four important dimensions.

First or third angle projection may be used but you must state on your paper which method you have adopted.

Time allowed – $1\frac{3}{4}$ hours.

Solution (page 150)

View (a)

The working position asked for is shown in the solution. However, the table can be rotated so that the T slots are in whatever position suits the purpose best.

View (b)

This view can be constructed by reproducing the given plan views of both table and base plate.

View (c)

A curve of intersection is required for the junction of the base with the 38 mm diameter piece. This will necessitate a portion of hidden detail relating to these two features being drawn in view (b).

ADVANCED PROBLEMS 21 TO 46
(Single Component and Assembly Drawings)

It is intended that the following problems consolidate the student's understanding of engineering drawing and also provide examples of examination papers in the subject from various institutes of higher education.

The examples alternate between single component and assembly drawings with the more time-consuming problems coming towards the end of the book.

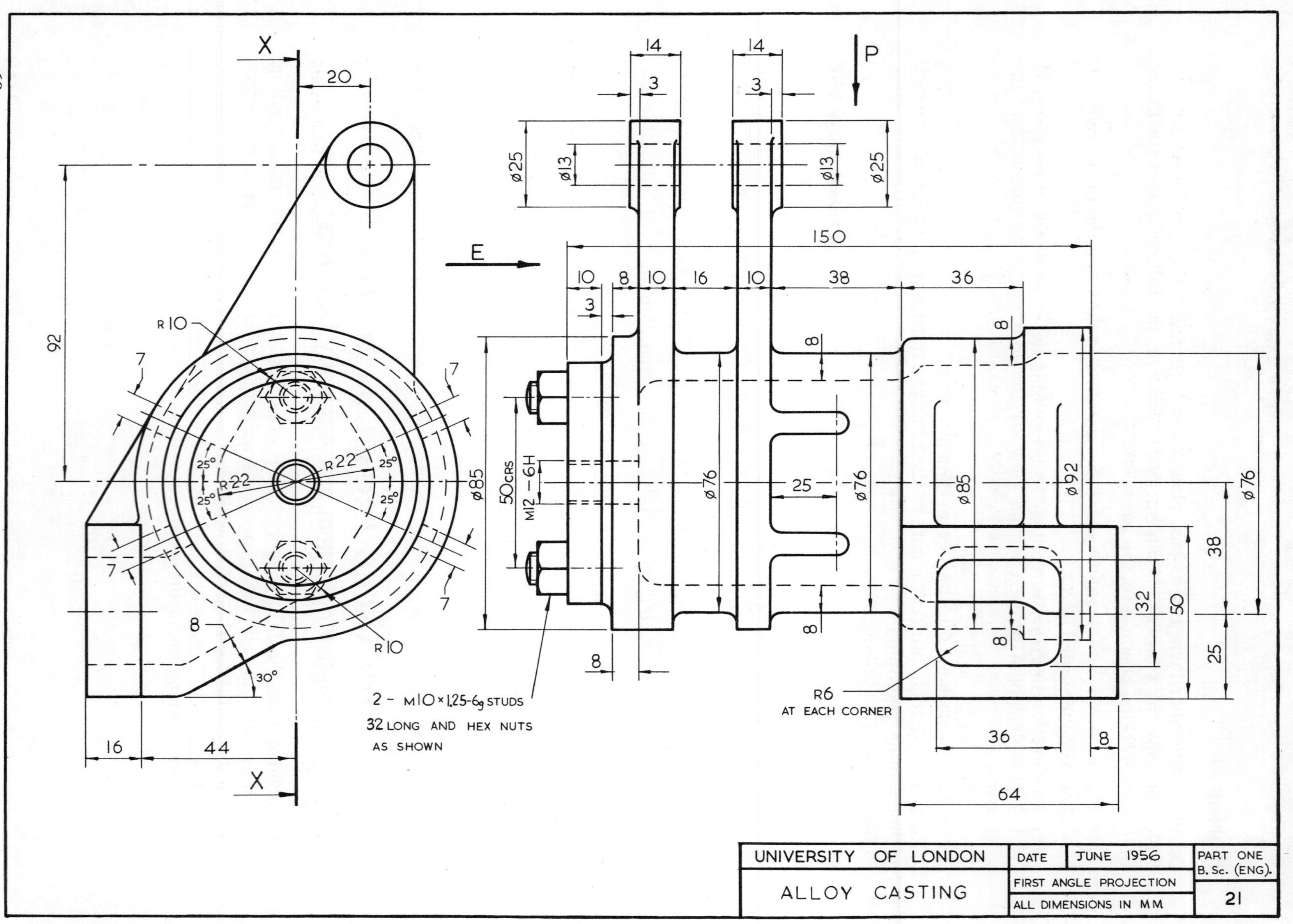
UNIVERSITY OF LONDON
DATE
JUNE 1956
PART ONE B. Sc. (ENG).
21
ALLOY CASTING
FIRST ANGLE PROJECTION
ALL DIMENSIONS IN MM
R6 AT EACH CORNER
2 - M10×1,25-6g STUDS
32 LONG AND HEX NUTS
AS SHOWN
50CRS
M12 - 6H
P
E
X
X
R10
R10
R22
R22
30°

Problem 21

Two elevations of an **alloy casting** fitted with a flange plate are shown in Figure 21. The given views are drawn in first angle projection.

Do not draw the views as shown but draw full size:

(a) A sectional elevation; the plane of the required section and the direction of the view are indicated at XX.

(b) A complete elevation to the right of and in projection with view (a) as seen from the direction of the arrow E.

(c) A complete plan in projection with view (a) and as seen from the direction of the arrow P.

Insert six dimensions on view (a), add the title Alloy Casting, and the scale.

The radii of small curves and fillets are to be chosen by the candidate. No hidden part lines are required on any of the views.

Time allowed – 2 hours.

Solution (page 151)

View (a)

The sectional elevation is viewed in a similar direction to the given front elevation, and consequently has a similar profile. The exception to this being the rectangular port which is in front of the cutting plane.

It must be assumed that the studs penetrate into the casting for a length equal to their diameter, i.e., 10 mm.

View (b)

Because it has an opposite viewing direction, view (b) is of the opposite hand to the given end elevation.

The two 10 mm thick ribs supporting the rectangular port blend with diameters that are of different sizes and must be shown accordingly.

View (c)

The main body of the casting is made up of concentric diameters and hence their plan view is similar to view (a).

It will be necessary to construct lightly one of the four 7 mm wide ribs in view (b) in order that their plan view may be drawn in view (c).

N.B. The projection method demanded in views (a) and (b) will result in first angle projection.

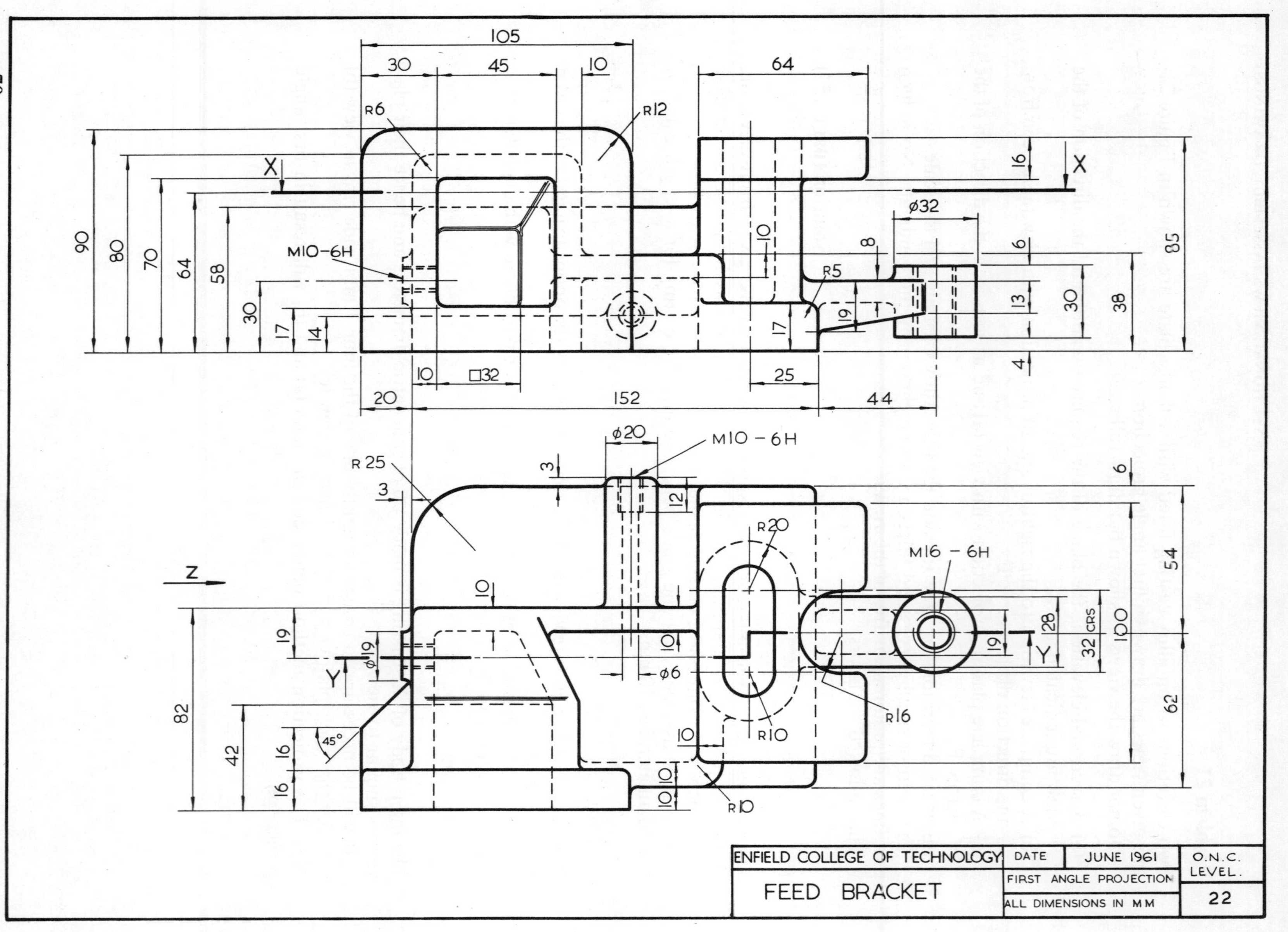
ENFIELD COLLEGE OF TECHNOLOGY
FEED BRACKET
DATE
JUNE 1961
O.N.C. LEVEL.
22
FIRST ANGLE PROJECTION
ALL DIMENSIONS IN MM
M16 - 6H
M10 - 6H
M10-6H
R16
R20
R10
R12
R6
R5
R 25
32 CRS
100
ø32
ø20
ø6
ø19
□32
45°
152
X
Y
Z

Problem 22

Figure 22 shows two views of a **feed bracket** for a milling machine.

Do not copy these views but draw full size, in third angle projection, the following views:

(a) A sectional view, the plane and direction of the section being indicated by the line XX.
(b) A sectional elevation, the plane and direction of the section being indicated by the line YY, and projected from view (a).
(c) An outside elevation, looking in the direction of arrow Z, and projected from view (a).

Insert six important dimensions and standard data.
Time allowed – 2 hours.

Solution (page 152)

View (a)

The profile is similar to that of the lower of the two given views but it is important to realise that the view cannot be completed until the outside elevation on arrow Z has been drawn. The sloping face, together with its wall thickness, must be drawn in the elevation on arrow Z and a line representing the plane XX inserted. It is then possible to project into Section XX that portion which is below the section plane.

View (b)

A close study of the hidden detail given is necessary to determine the features cut by the offset section plane YY. The outside elevation on Z, and a line representing the plane YY, are necessary for similar reasons to those for view (a).

View (c)

Construct the nearest features first. That is, the front faceplate on the views right hand together with the small bracing web. (q) and the bracket's base plate (p). The hollow box with its sloping roof must be drawn with part of its wall thickness shown by hidden part lines for projections into views (a) and (b).

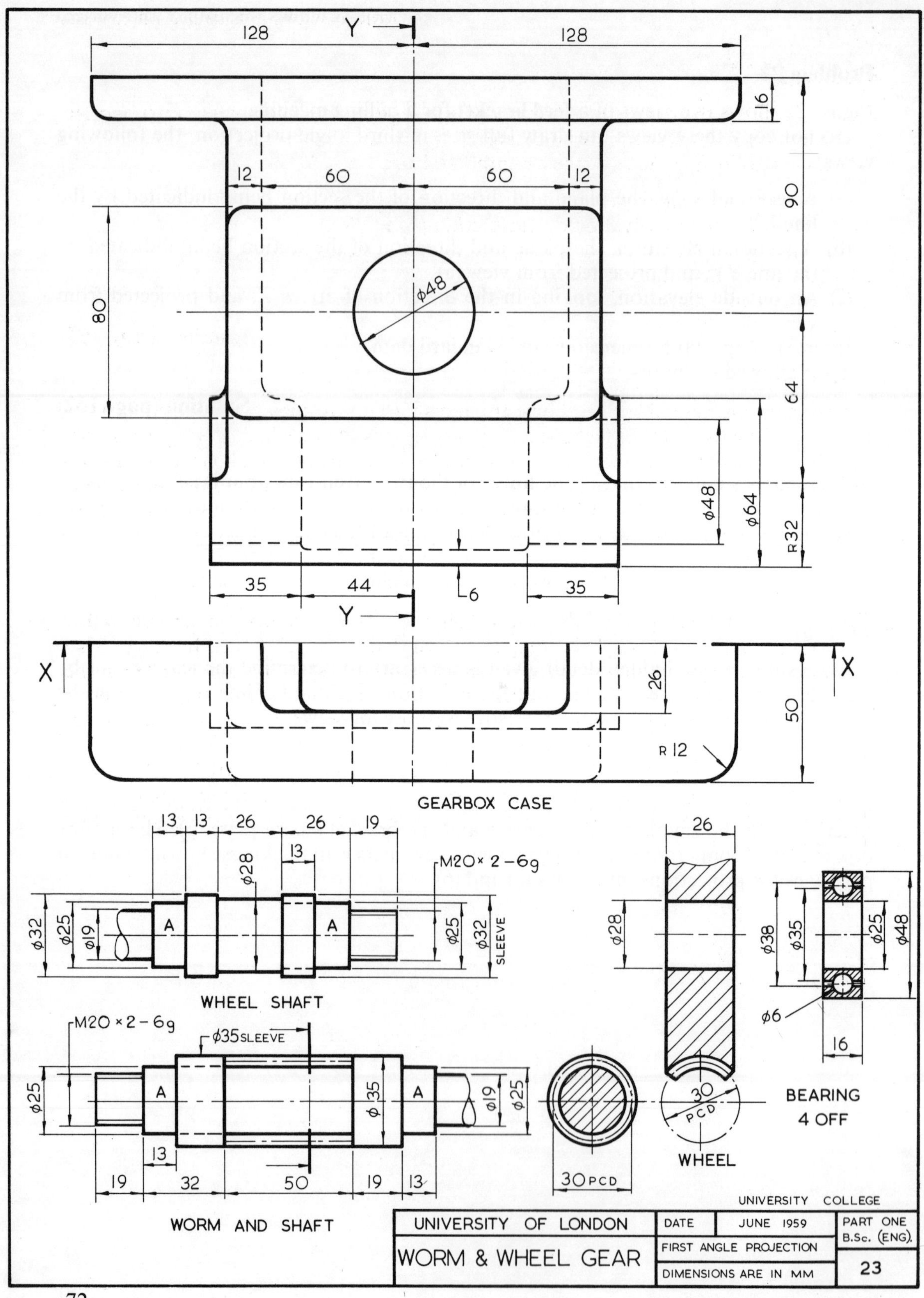
128
Y
128
16
12
60
60
12
90
80
φ48
64
φ48
φ64
R32
35
44
6
35
Y
X
26
50
X
R 12
GEARBOX CASE
13
13
26
26
19
13
φ28
M20× 2 –6g
φ32
φ25
φ19
A
A
φ25
φ32
SLEEVE
WHEEL SHAFT
26
φ28
φ38
φ35
φ25
φ48
φ6
16
M20 × 2 – 6g
φ35SLEEVE
φ25
A
φ 35
A
φ19
φ25
13
30
PCD
BEARING
4 OFF
WHEEL
19
32
50
19
13
30PCD
WORM AND SHAFT
UNIVERSITY COLLEGE
UNIVERSITY OF LONDON
DATE
JUNE 1959
PART ONE
B.Sc. (ENG).
WORM & WHEEL GEAR
FIRST ANGLE PROJECTION
DIMENSIONS ARE IN MM
23

Problem 23

Figure 23 shows the details of a **worm and wheel gear**. The two shafts are supported by the ball bearings at the places indicated.

Draw the following views of the assembly full size:

(a) A front sectional elevation on the section plane XX.
(b) An end sectional elevation on the section plane YY.
Hidden edges need not be shown as dotted lines in either view.
Estimate any missing dimensions.
Time allowed – 2 hours.

Solution (page 153)

View (a)

The bearings are assembled on the worm and shaft at positions A. The distance from the centre of the worm to the furthest face of the sleeve is 44 mm which is also the distance from the centre of the gearbox case to the inside face of the bearing housings. This will position a bearing flush with the inside face of the bearing housing and, for correct assembly, the other bearing should be assembled in a similar position. Either side of the gearbox case may be chosen for the shaft extension.

View (b)

The bearings are assembled on the wheel shaft at positions A and a similar case to that for view (a) can be made for positioning the wheel shaft bearings in their housings. That is, the centre of the 28 mm diameter portion of the wheel shaft is 26 mm from the furthest face of the integral collar, which coincides with the distance from the centre of the gearbox case to the inside face of the bearing housings. Similarly, either side of the case may be chosen for the shaft extension.

The solution on page 153 illustrates the bearings pictorially to help the student to comprehend the role fulfilled by a bearing. However, the student is reminded that BS 308 includes a conventional representation for bearings and it is this representation that should normally be used.

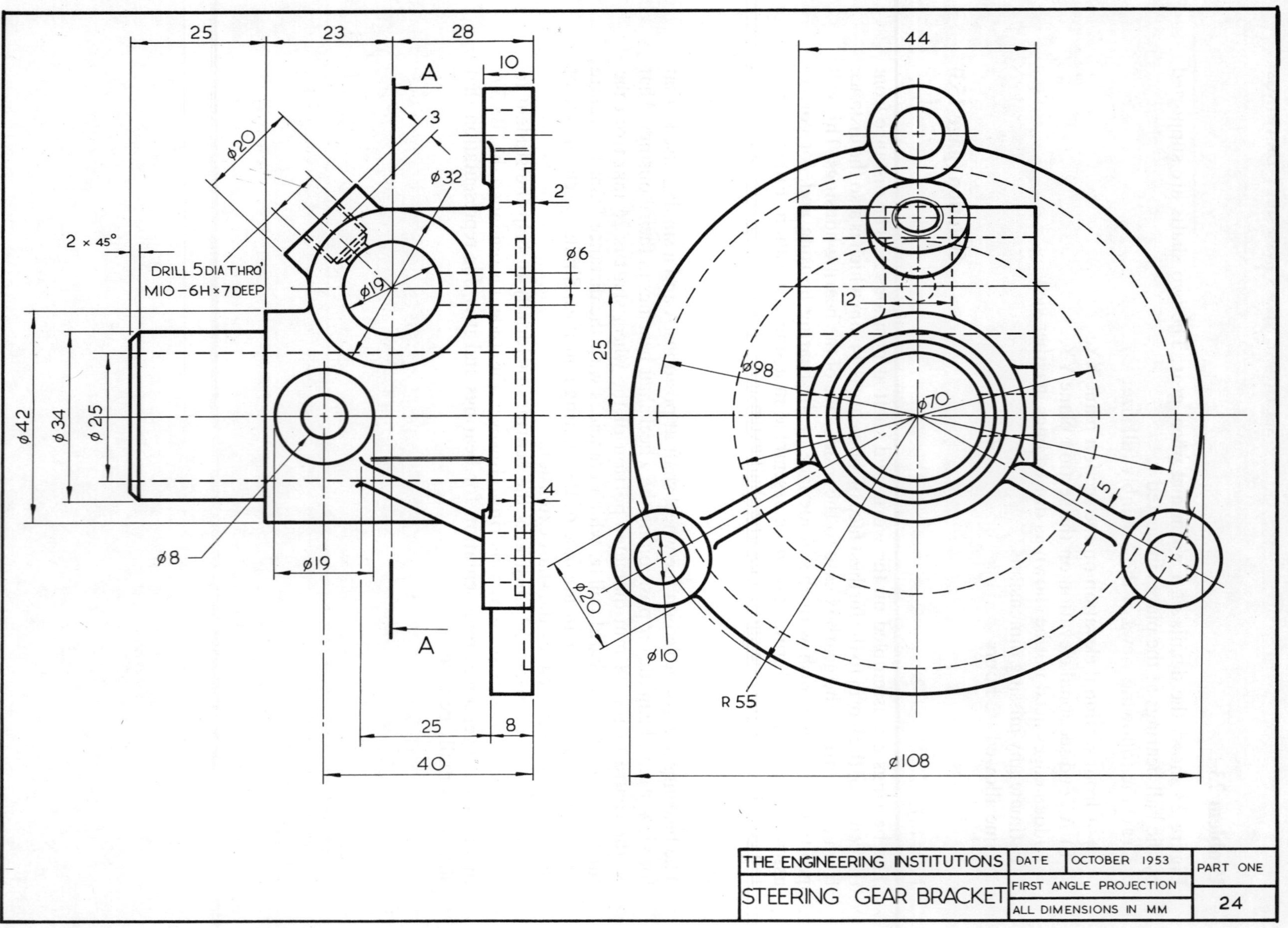
25
23
28
10
A
3
⌀20
⌀32
2
2 × 45°
DRILL 5 DIA THRO'
M10 – 6H × 7 DEEP
⌀6
⌀19
25
⌀42
⌀34
⌀25
4
⌀8
⌀19
A
25
8
40
44
12
⌀98
⌀70
5
⌀20
⌀10
R 55
⌀108
THE ENGINEERING INSTITUTIONS
DATE
OCTOBER 1953
PART ONE
STEERING GEAR BRACKET
FIRST ANGLE PROJECTION
ALL DIMENSIONS IN MM
24

Problem 24

Figure 24 shows details of a **steering gear bracket** in gunmetal.

Draw the following views full size:
(a) The sectional elevation.
(b) The sectional end elevation on AA.
(c) The plan, the lower half in section on the principal section plane.

Use first or third angle projection. Dimension the view (a) only, and add dotted lines and section lines in all views. Candidates are expected to decide any dimensions omitted from the drawing provided.

Time allowed – 2 hours.

Solution (page 154)

View (a)

The required sectional view is the result of a cutting plane passing through the centre line of the component.

Before the lower web can be drawn – the majority being in hidden detail – it will be necessary to construct lightly one of them in view (b).

Note that an angle of 30° has been chosen for the cross hatch-lines to prevent them from being parallel or perpendicular to the outline of the 20 mm dia boss.

View (b)

Only a very small amount of thc componcnt is visible behind the sectional plane because of the viewing direction. The portion of the 20 mm dia boss that is visible and its completion in hidden detail will require an ellipse to be constructed.

The length of the webs should be lightly drawn in order that they can be projected for their construction in views (a) and (c).

With the section plane drawn in view (a) the sectioned length of the webs can then be projected back into view (b).

View (c)

The top half is an outside view when viewed from above the component and the lower half is a section on plane BB since the component is generally symmetrical about this centre line.

The 20 mm dia boss will again appear as a semi-ellipse and a curve of intersection (intersecting cylinders) will be necessary for this boss and also the side boss of the same dia.

The shape descriptions of the webs can be constructed as described for view (b).

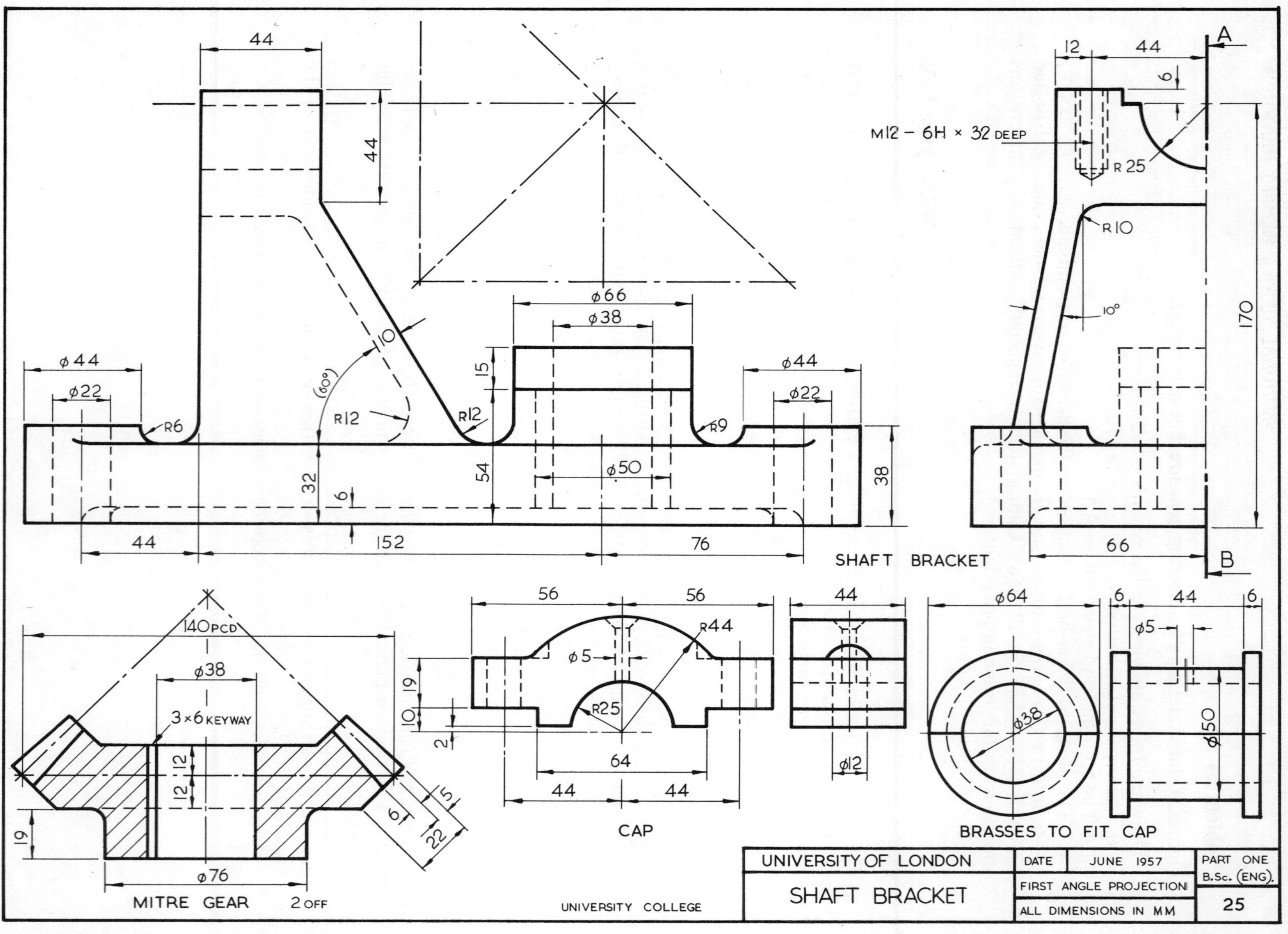
M12 – 6H × 32 DEEP
R 25
R 10
10°
170
A
B
66
12
44
6
φ66
φ38
φ44
φ22
15
(60°)
R12
R6
R9
10
φ50
32
54
6
38
44
152
76
SHAFT BRACKET
140 PCD
φ38
3×6 KEYWAY
12
12
19
5
6
22
φ76
MITRE GEAR 2 OFF
56
56
R44
φ5
R25
19
10
2
64
44
44
CAP
44
φ12
φ64
φ38
6
44
6
φ5
φ50
BRASSES TO FIT CAP
UNIVERSITY COLLEGE
UNIVERSITY OF LONDON
SHAFT BRACKET
DATE JUNE 1957
FIRST ANGLE PROJECTION
ALL DIMENSIONS IN MM
PART ONE B.Sc. (ENG).
25

Problem 25

Figure 25 shows the details of a **shaft bracket**.

Draw the following views of the assembly, full size:
(a) A front sectional elevation on the plane AB.
(b) One half of the plan view.
Hidden edges need not be shown as dotted lines. In view (a), show a part of a 38 mm dia shaft, to which is fastened a mitre gear with a suitable gib head key. The second gear should be shown but the shaft and key may be omitted.
In view (b) the gears may be omitted completely.
Time allowed – 2 hours.

Solution (page 155)

View (a)

The brasses, which are in the form of a split tube, are used to facilitate assembly and maintenance. By removing the cap and brasses the shaft and bevel (or mitre) gear assembly can be dismantled without withdrawing the shaft axially. The other gear runs in a bush which is already assembled in the given views of the shaft bracket. Reference to BS 308 should be made for the conventional method of drawing bevel gears.

View (b)

A stud and hexagonal nut has been chosen to secure the cap to the bracket. This will prevent damage to the thread in the bracket due to the cap being removed frequently for maintenance, etc. The length of the line representing the edge of the 10° sloping face in plan view, terminating at Q, will necessitate that a portion of the given end elevation be drawn lightly.

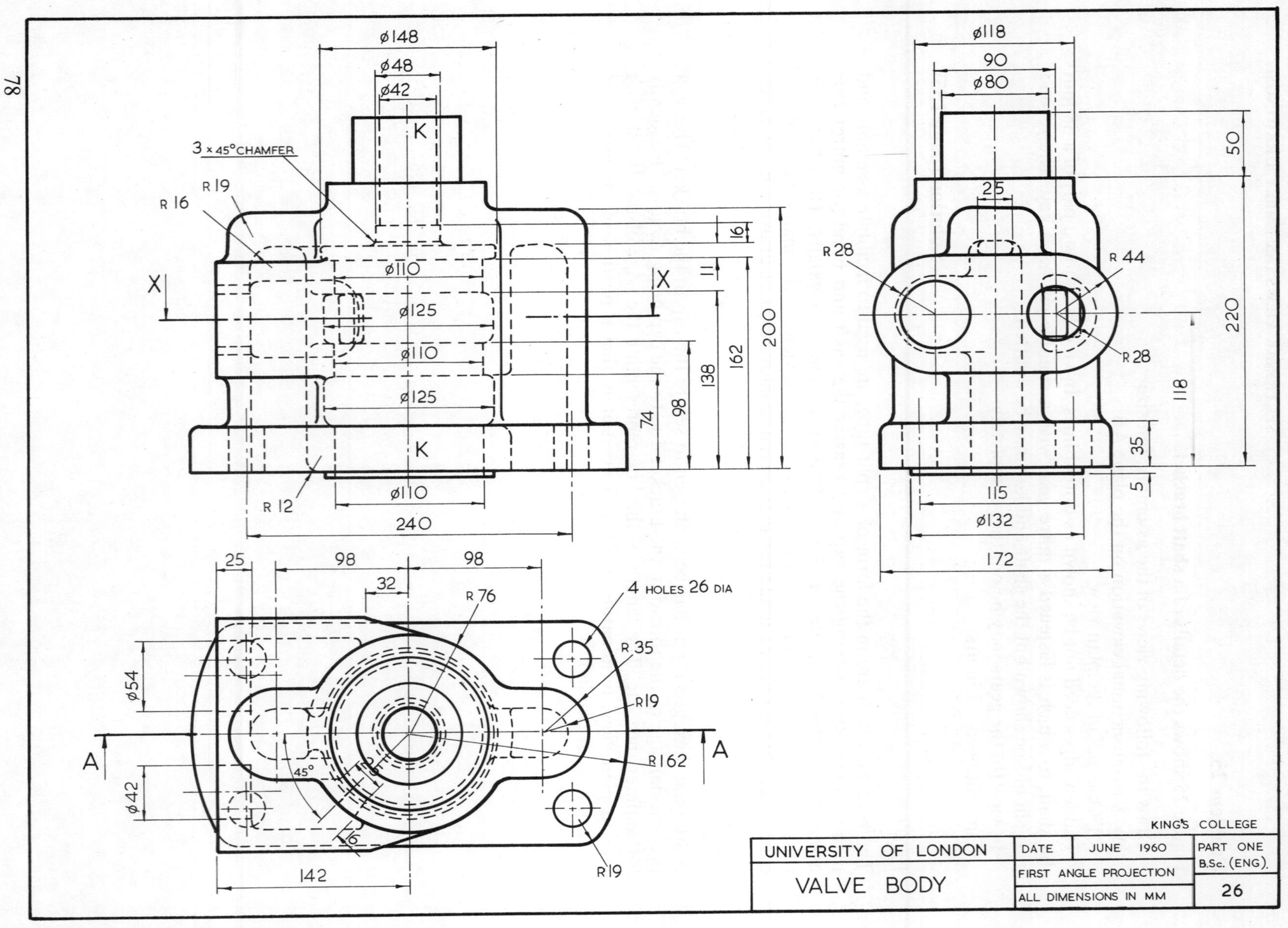
ø148
ø48
ø42
K
3 × 45° CHAMFER
R 19
R 16
X
ø110
ø125
ø110
ø125
K
R 12
ø110
240
16
11
74
98
138
162
200
ø118
90
ø80
50
25
R 28
R 44
R 28
220
118
35
5
115
ø132
172
25
98
98
32
R 76
4 HOLES 26 DIA
R 35
R19
R162
A
ø54
ø42
45°
16
142
R19
KING'S COLLEGE
UNIVERSITY OF LONDON
VALVE BODY
DATE JUNE 1960
FIRST ANGLE PROJECTION
ALL DIMENSIONS IN MM
PART ONE B.Sc. (ENG).
26

Problem 26

Figure 26 shows three views of a **valve body** which is part of the governing device for a steam turbine.

(a) Draw a sectioned view on AA.
(b) Draw a sectioned view in correct projection with view (a), the section being indicated by arrows XX.
The centre line KK should be parallel to the longer edges of your drawing paper.
Use a $\frac{3}{4}$ scale and first angle projection.
No hidden detail is required.
(c) Add seven important dimensions, the system of projection and the title Valve Body.

A few dimensions have been omitted and you are to assume reasonable values for them.

Time allowed – 2 hours.

Solution (page 156)

View (a)

A similar profile to the given front elevation is necessary and a close study of the hidden detail to determine how much of the hidden detail refers to features cut by plane AA. In fact, all but the small quantity relating to the 25 mm slot, positioned at 45° in the plan view, must be drawn, and as full lines.

The 'necking', shown in hidden detail on the horizontal centre line in the plan view, occurs at a distance of 162 mm up from the base of the valve and for a thickness of 11 mm.

View (b)

The left hand half of this view possesses several problems and these will be discussed separately:

(i) As can be seen from the given end elevation, the plane XX cuts through holes of dissimilar size.
(ii) The left hand of these is clear and provides access to a shelf which is 90 mm up from the base of the valve.
(iii) The right hand hole which is blind for a small portion provides access to a cavity, which is hollow through to the base, and is in line with the 25 mm slot.

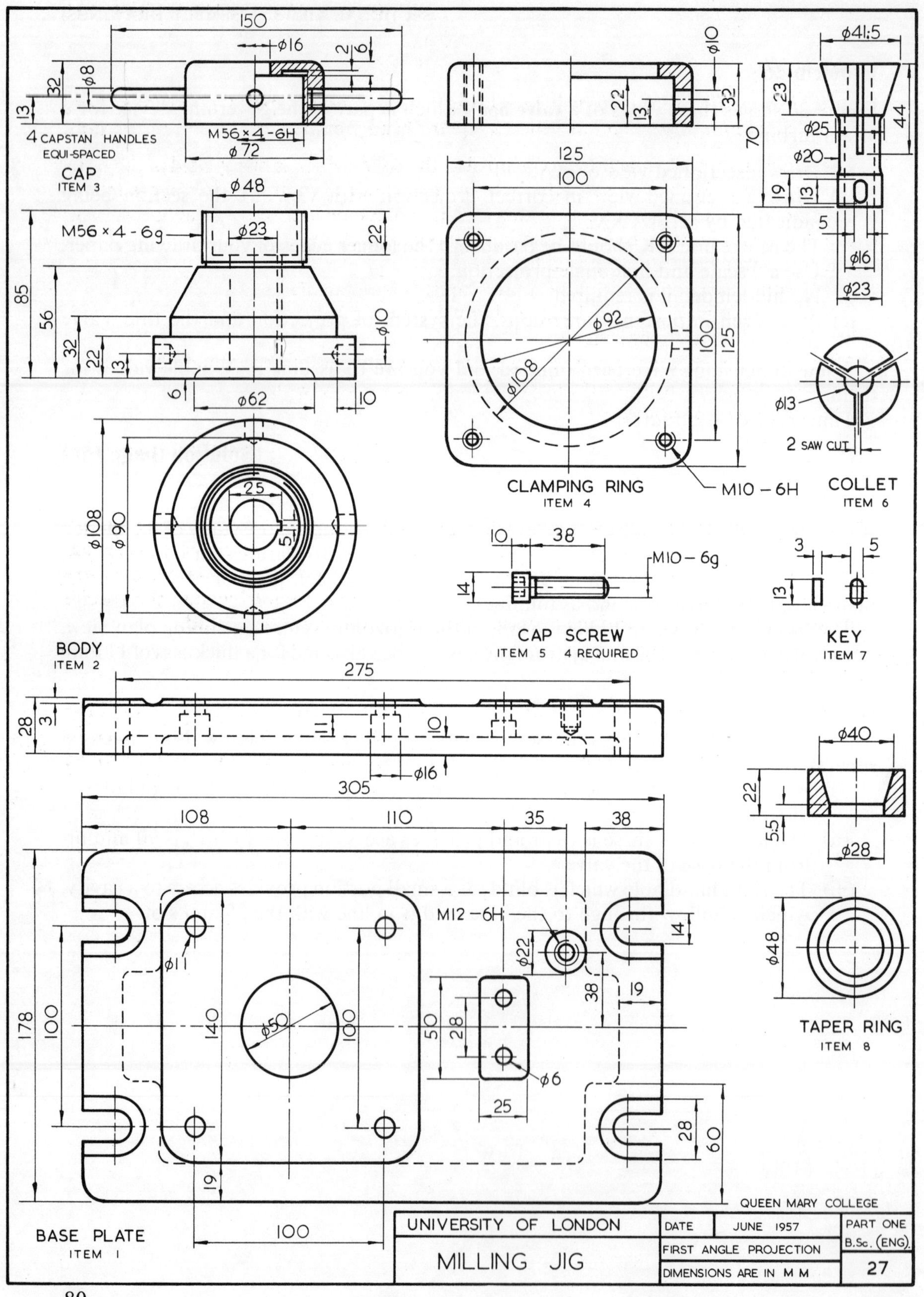
4 CAPSTAN HANDLES EQUI-SPACED
M56 × 4 - 6H
CAP
ITEM 3
M56 × 4 - 6g
BODY
ITEM 2
CLAMPING RING
ITEM 4
M10 - 6H
COLLET
ITEM 6
2 SAW CUT
M10 - 6g
CAP SCREW
ITEM 5 4 REQUIRED
KEY
ITEM 7
M12 - 6H
TAPER RING
ITEM 8
BASE PLATE
ITEM 1
QUEEN MARY COLLEGE
UNIVERSITY OF LONDON
MILLING JIG
DATE JUNE 1957
FIRST ANGLE PROJECTION
DIMENSIONS ARE IN M M
PART ONE
B.Sc. (ENG).
27

Problem 27

Figure 27 shows details of a **milling jig**. The jig is used to hold a special bolt (not shown) during a machining operation in which a square head is milled on the end of the bolt.

The component to be machined is gripped in the collet which is firmly held in the body of the jig. The body is secured to the base by means of the clamping ring but is free to rotate into one of four possible positions in which it can be locked by a pin (not shown) which engages in mating holes drilled in the body and clamping ring.

The following views are required full size:

(a) A sectional front elevation of the assembled components.

(b) An outside plan view of the assembly with the cap removed.

No dotted lines are required.

All missing dimensions and small radii are to be assessed by the candidate.

A few leading dimensions may be inserted on the views drawn.

Time allowed – $2\frac{1}{2}$ hours.

Solution (page 157)

View (a)

The special bolt which is to be machined is gripped in the collet by the 'wedge' effect of the collet's outer surface when forced downwards by the cap. The collet is prevented from rotating by the small key. The four 10 mm diameter holes in the body, together with the single 10 mm diameter hole in the clamping ring, are for locating the special bolt in each of the four positions necessary for machining its square head.

View (b)

The given plan view of the base plate can be reproduced together with plan views of the collet, taper ring, body and clamping ring. The cap screws do not protrude through to the top surface of the clamping ring and as a consequence tapped holes are drawn in their four positions.

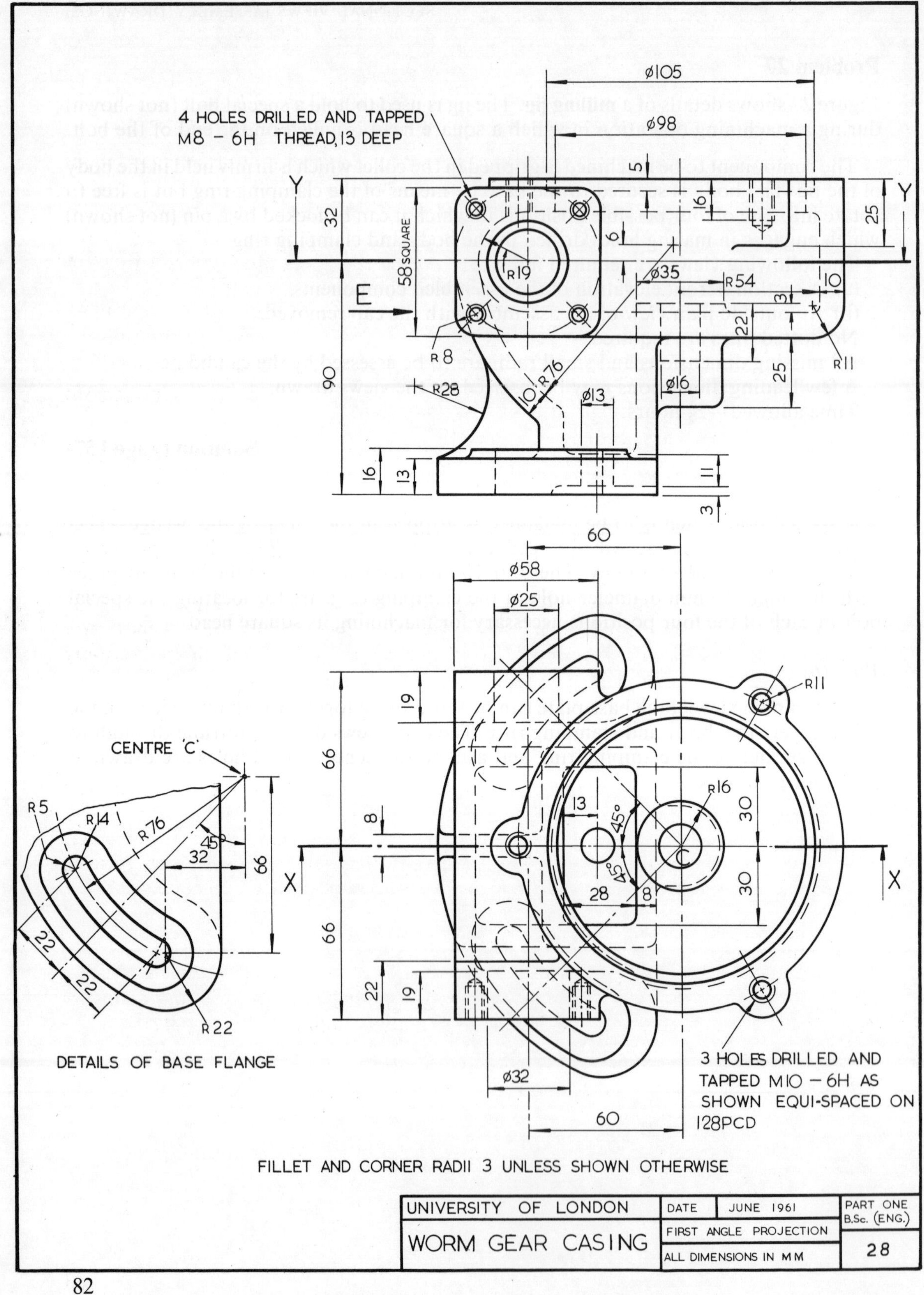
4 HOLES DRILLED AND TAPPED
M8 - 6H THREAD, 13 DEEP
ø105
ø98
58 SQUARE
R19
ø35
R54
R8
R28
R76
ø16
ø13
R11
CENTRE 'C'
R5
R14
R76
45°
32
66
22
22
R22
DETAILS OF BASE FLANGE
60
ø58
ø25
R11
R16
13
45°
28
8
30
30
X
X
Y
Y
E
ø32
60
3 HOLES DRILLED AND
TAPPED M10 - 6H AS
SHOWN EQUI-SPACED ON
128PCD
FILLET AND CORNER RADII 3 UNLESS SHOWN OTHERWISE
UNIVERSITY OF LONDON
DATE JUNE 1961
PART ONE B.Sc. (ENG.)
WORM GEAR CASING
FIRST ANGLE PROJECTION
ALL DIMENSIONS IN MM
28

Problem 28

Figure 28 shows two orthographic views of a **worm gear casing**.
Do not copy the views as shown but draw full size the following views:

(a) A sectional elevation taken on the plane XX, the view required being that as seen when looking in the direction of the arrows.
(b) An outside elevation as seen when looking in the direction of the arrow E.
(c) A part of the sectional plan, the plane of the section and the direction of the required view being indicated YY. The view required is of that portion which is shown above the centre line XX in the plan view.

No hidden edges are to be shown in any of the views. Draw in the lower right hand corner of the drawing paper a title block 100 mm by 60 mm and insert the title and scale.
Time allowed – 2 hours.

Solution (page 158)

A pictorial view, illustrating the effect of both the sectional planes, has been included with the required solution as an aid to visualisation. The web has been sectioned for clarity.

View (a)

As the pictorial view shows, the square boss, being part of the portion that has been removed, is not seen and view (a) can be built up by constructing the remainder of the given front elevation with the hidden detail lines drawn as full lines. Note that the centre web has not been sectioned. The centre of the 76 mm rad can be determined by describing an arc of 57 mm from the centre of the 25 mm and 32 mm diameter holes to cut a line 6 mm above centre line YY.

View (b)

The pictorial view has been sectioned at plane YY and illustrates the half that is required in view (b). It should be noted that the lugs for the M10 – 6H tapped holes are above the cutting plane and hence are not included in view (b). As can be seen in the given plan view it would appear that the two corners of the main pillar, of width 76 mm, join the base at the base flanges which will prevent the top surface of the flanges being machined completely.

View (c)

The position of intersection of the main 76 mm wide pillar with the 58 mm diameter bosses can be projected from view (a), as can the intersection of web and pillar.

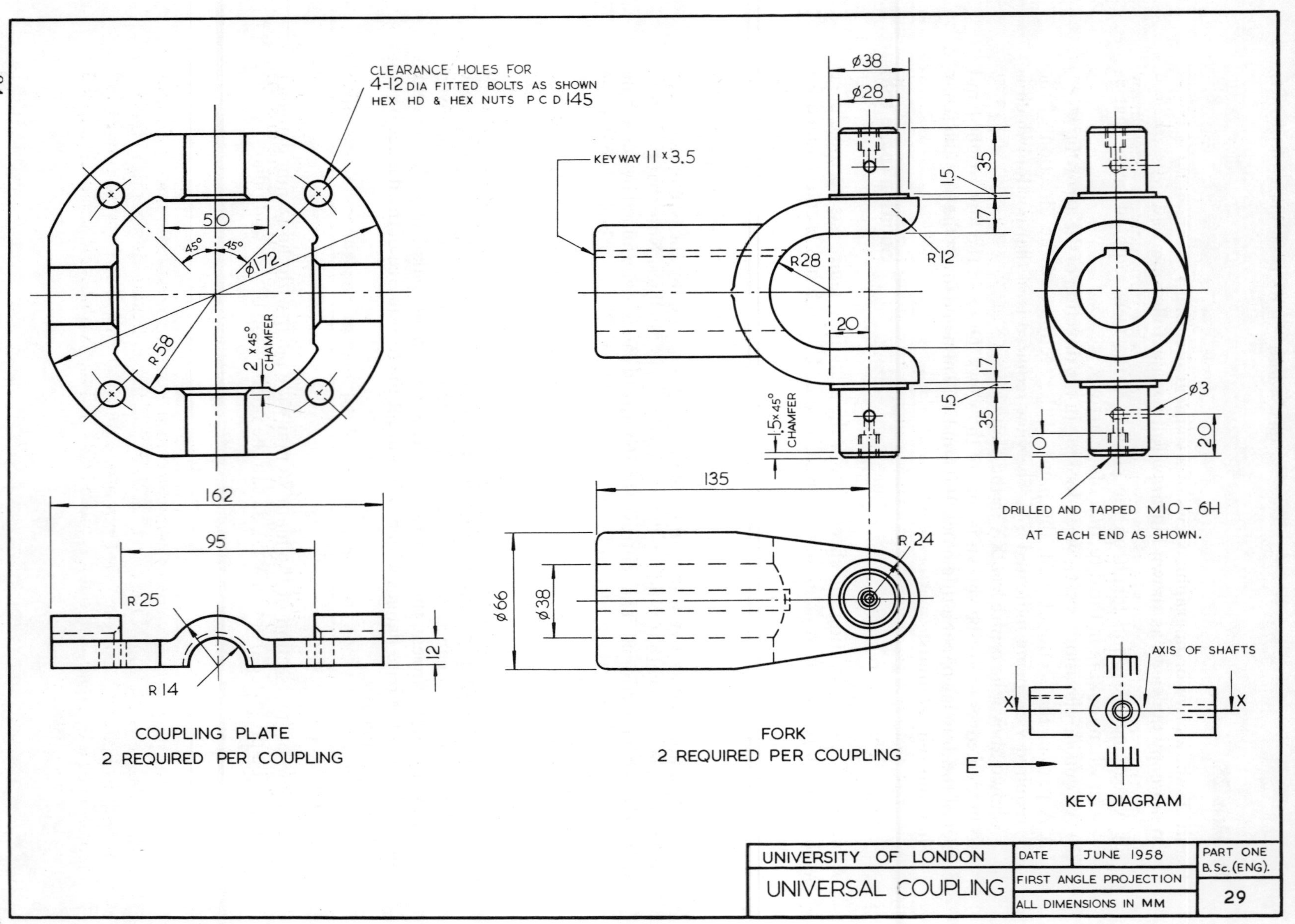
CLEARANCE HOLES FOR
4-12 DIA FITTED BOLTS AS SHOWN
HEX HD & HEX NUTS P C D 145
50
45°
45°
ø172
R 58
2 x 45° CHAMFER
162
95
R 25
R 14
12
COUPLING PLATE
2 REQUIRED PER COUPLING
KEY WAY 11 x 3.5
ø38
ø28
35
1.5
17
R12
R28
20
17
1.5
35
1.5 x 45° CHAMFER
135
ø66
ø38
R 24
FORK
2 REQUIRED PER COUPLING
ø3
10
20
DRILLED AND TAPPED M10 – 6H
AT EACH END AS SHOWN.
AXIS OF SHAFTS
X
X
E
KEY DIAGRAM
UNIVERSITY OF LONDON
UNIVERSAL COUPLING
DATE
JUNE 1958
FIRST ANGLE PROJECTION
ALL DIMENSIONS IN MM
PART ONE
B.Sc. (ENG).
29

Problem 29

The details of a **universal coupling** are shown in Figure 29. Two similar coupling plates and two similar forks are required to form the complete coupling. The two forks are held with their ends between the two coupling plates, which are jointed together by means of four 12 mm dia bolts and nuts.

Do not draw the separate parts as shown, but produce, full size, the following views of the assembled coupling:

(a) An outside elevation, showing the axis of the two shafts in a horizontal position, and conforming to the general position indicated in the key diagram shown in the figure.

(b) An end elevation, looking in the direction indicated by the arrow E. This view is to be placed to the right of view (a).

(c) A half sectional plan. The plane of the section is to contain the axis X – X of the shafts. The view required is that which appears above the axis of the shafts, and is to be placed under view (a).

Insert two dimensions only in each view. Draw a title block 100 mm by 60 mm in the bottom right hand corner of your drawing and insert the title and scale.

No hidden part lines are required in any of the views. Only one bolt and nut is to be shown, and this must be correctly projected in views (a) and (b).

Solution (page 159)

A pictorial view, which has been partially sectioned, has been included with the required solution to help visualise the arrangement of the assembled parts.

View (a)

The given views of the fork are arranged in similar positions to those of the left hand fork in the required views of the assembled coupling.

The right hand fork is assembled in a position disposed at 90° to the left hand fork and both clamped by the two coupling plates. For the construction of the curve of intersection between ‘l’ and ‘m’, view (c) must first be drawn and a procedure adopted as for intersecting cylinders.

View (b)

The solution requires that a view on arrow E is positioned to the right of view (a) which is the case for first angle projection.

Shaded lines have been shown to indicate the junction of the housings for the fork pivot pins and the parent plates. Note that the arc across the keyway P represents the bore of the hidden fork.

View (c)

Similarly, the position for view (c) demanded in the question results in first angle projection. The construction of the curve ‘qr’ is as for perpendicular intersecting cylinders.

It could be argued that the pivot pin of the fork is a part that drawing convention recommends is not sectioned but it is more likely that the examiners intended the tapped hole to be shown in section.

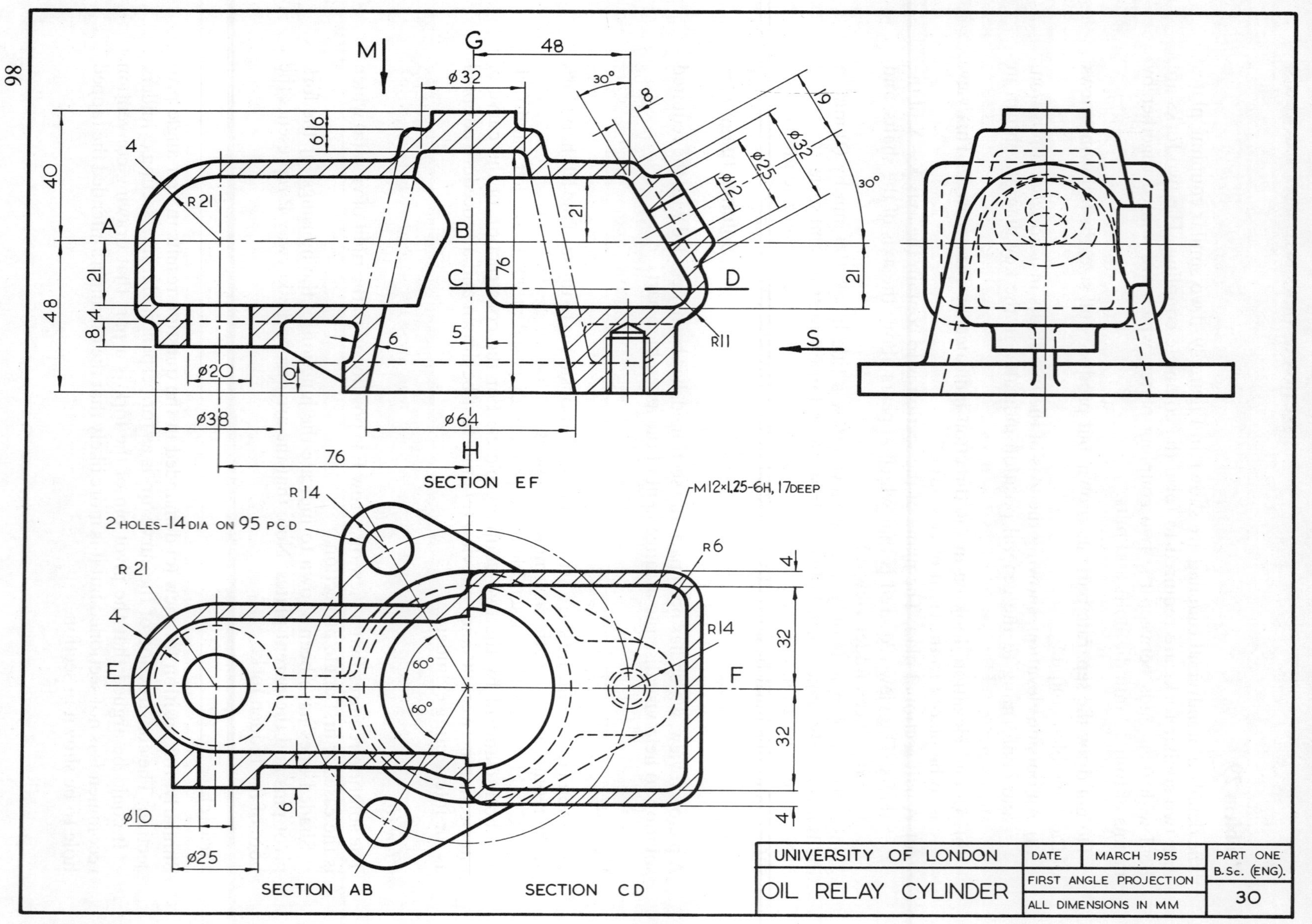
SECTION EF
SECTION AB
SECTION CD
2 HOLES-14 DIA ON 95 PCD
M12×1.25-6H, 17DEEP
UNIVERSITY OF LONDON
DATE MARCH 1955
PART ONE B.Sc. (ENG).
OIL RELAY CYLINDER
FIRST ANGLE PROJECTION
ALL DIMENSIONS IN MM
30

Problem 30

Views of the casting for an **oil relay cylinder** are shown in Figure 30.

Draw the following views, full size:
(a) An outside elevation. This is to replace the sectional view EF.
(b) A plan view, to be placed under view (a) and to be in projection with that view. The view is to be that seen when looking in the direction of the arrow M.
(c) A sectional side elevation, to be placed to the left of view (a) and to be in projection with that view. The plane of the required section is indicated by the line GH. The view is to be that seen when looking in the direction of the arrow S.
No hidden edges are required in any view.
Add six important dimensions on one view, together with the title and the scale.
Time allowed – $1\frac{3}{4}$ hours.

Solution (page 160)

View (a)

The profile of the given section EF can be repeated for view (a) together with a number of curves of intersection.

The centre of the body is the frustum of a cone, and for the curve of intersection between 'l' and 'm' the construction is as for the intersection of cylinder and cone. The positioning of line 'qr' requires that the 28 mm wide piece is drawn lightly in view (b).

View (b)

The viewing direction and position results in first angle projection. The profile of the given view involving sections on AB and CD can be repeated for view (b). The plan view of line 'lm' should have been constructed in conjunction with its construction in view (a). The tapered boss on the right hand sloping face can be projected from view (a) or one of the conventional methods for constructing an ellipse may be used. Remember that a line parallel to an ellipse is not an ellipse.

View (c)

As for view (b), the viewing direction and position results in first angle projection. A line representing the cutting plane XX must be drawn in view (b) in order that the width of the triangular base plate, at the cutting plane, can be determined.

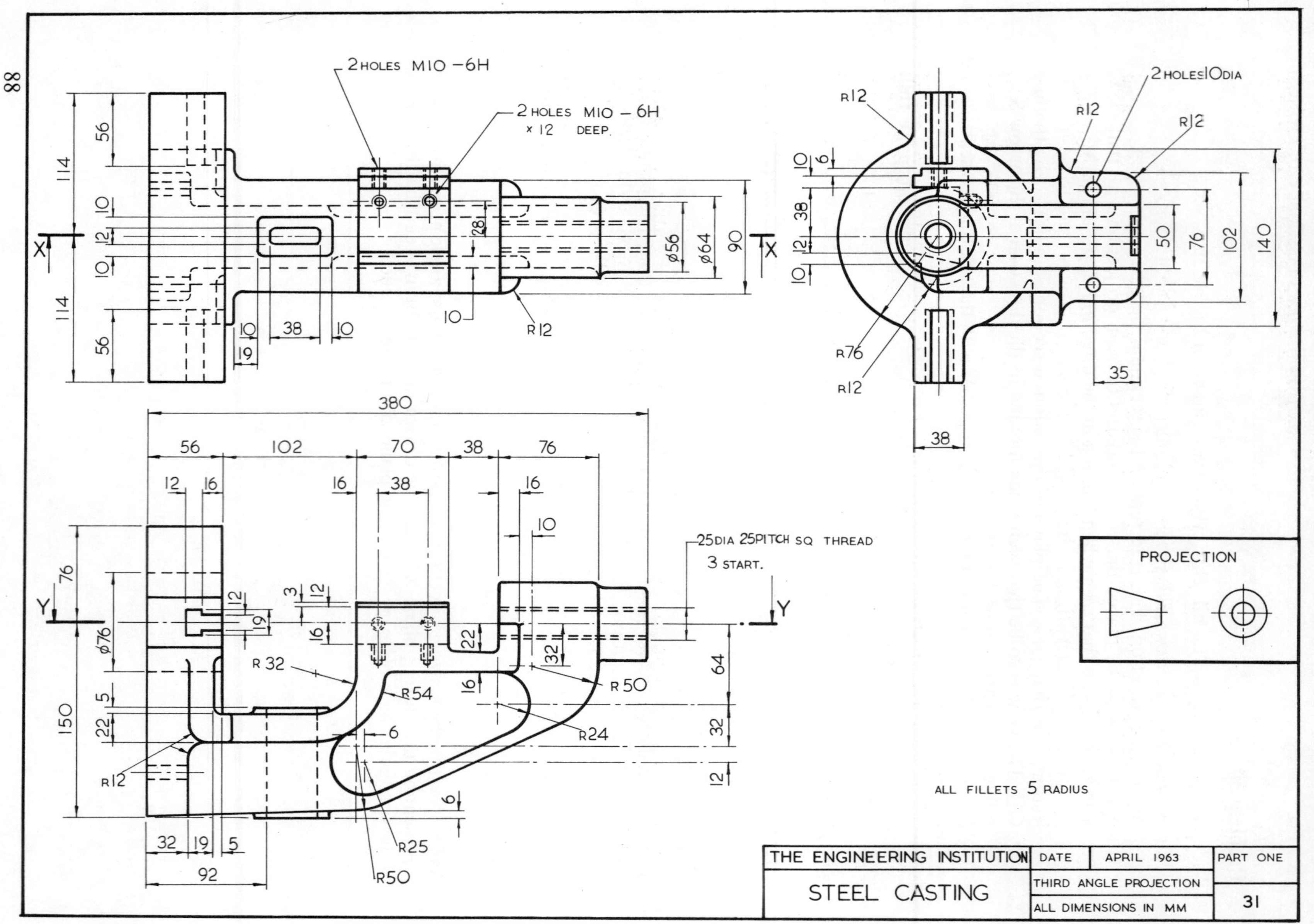
2 HOLES M10 – 6H
2 HOLES M10 – 6H × 12 DEEP.
2 HOLES 10 DIA
25 DIA 25 PITCH SQ THREAD 3 START.
PROJECTION
ALL FILLETS 5 RADIUS
THE ENGINEERING INSTITUTION
DATE
APRIL 1963
PART ONE
STEEL CASTING
THIRD ANGLE PROJECTION
ALL DIMENSIONS IN MM
31

Problem 31

Figure 31 shows details of a **steel casting**.

Draw half full size, using either method of projection, the following views:

(a) A sectional elevation, the plane and direction of the required view being indicated by the line XX.
(b) A sectional plan, the plane and the direction of the required view being indicated by the line YY.

Time allowed – $1\frac{3}{4}$ hours.

Solution (page 161)

View (a)

Section XX has a similar profile to that of the lower of the given elevations. As an aid to the visualisation of section XX a line representing the cutting plane should be extended through to the given end elevation. The section plane can be seen to pass through the centre of the boss containing the square threaded hole and also through the centre of the web dividing the two shaped recesses (marked P on the solution). In view of the considerable thickness of the web – 30 mm – it could perhaps be argued that it is too thick to be called a web and hence should be cross-hatched. However, the author believes the solution shown on page 161 to be a better interpretation.

View (b)

Section YY also has a similar profile to a given elevation with the cutting plane again passing through the boss containing the square threaded hole. A line representing the cutting plane drawn on the given end elevation will show that the plane passes through T slots on each side of the casting and these are shown free from cross hatching on view (b). A scrap section of the sloping face will be necessary in order that line 'ab' can be determined.

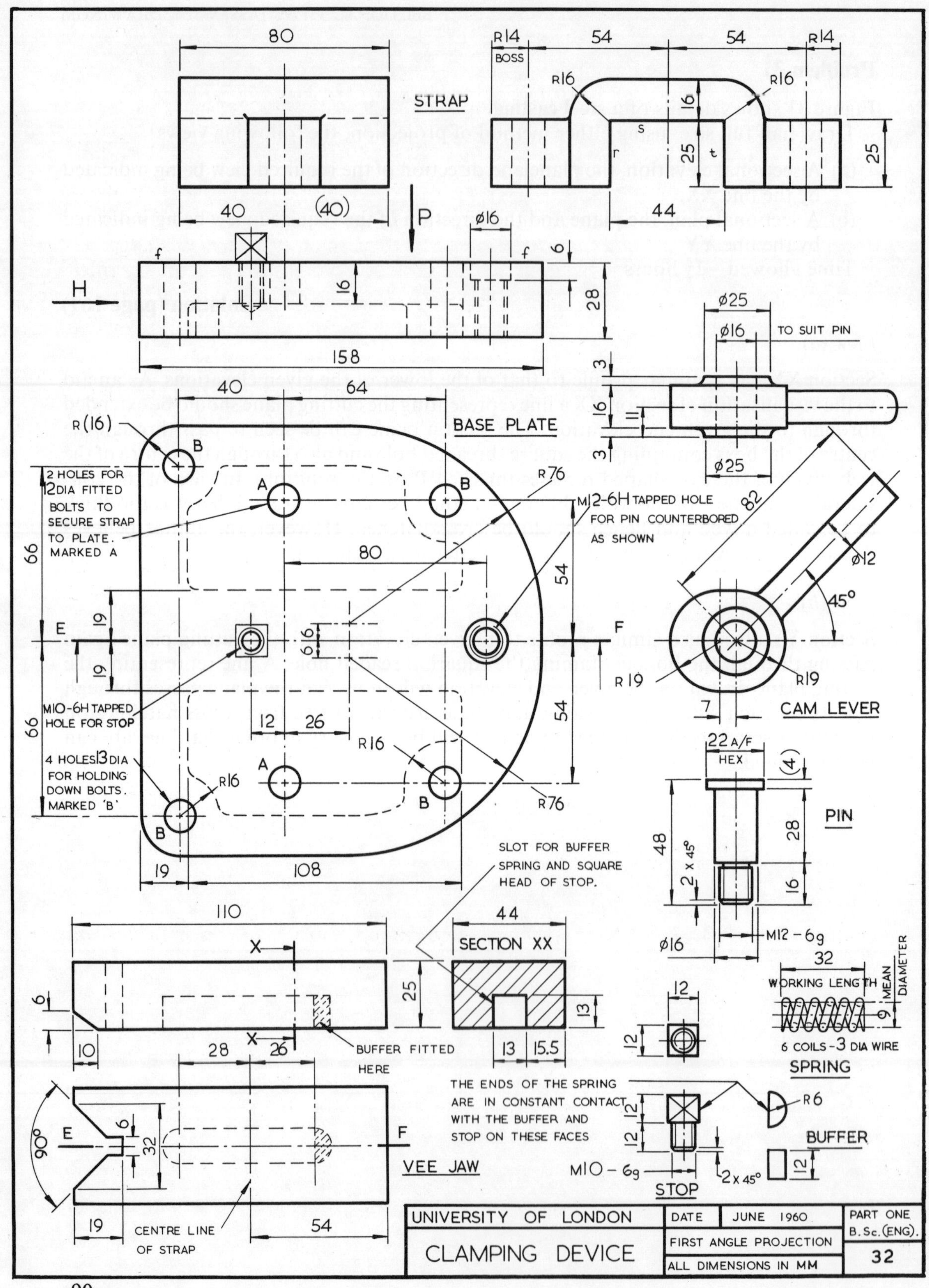
STRAP
BOSS
P
H
BASE PLATE
2 HOLES FOR 12 DIA FITTED BOLTS TO SECURE STRAP TO PLATE. MARKED A
M12-6H TAPPED HOLE FOR PIN COUNTERBORED AS SHOWN
MIO-6H TAPPED HOLE FOR STOP
4 HOLES 13 DIA FOR HOLDING DOWN BOLTS. MARKED 'B'
TO SUIT PIN
CAM LEVER
22 A/F HEX
PIN
M12 - 6g
SLOT FOR BUFFER SPRING AND SQUARE HEAD OF STOP.
SECTION XX
BUFFER FITTED HERE
WORKING LENGTH
MEAN DIAMETER
6 COILS - 3 DIA WIRE
SPRING
THE ENDS OF THE SPRING ARE IN CONSTANT CONTACT WITH THE BUFFER AND STOP ON THESE FACES
BUFFER
M10 - 6g
2 x 45°
STOP
VEE JAW
CENTRE LINE OF STRAP
UNIVERSITY OF LONDON
DATE
JUNE 1960
PART ONE B.Sc.(ENG).
CLAMPING DEVICE
FIRST ANGLE PROJECTION
ALL DIMENSIONS IN MM
32

Problem 32

The details of a **clamping device** are shown in first angle projection in Figure 32.

The locking piece labelled 'vee jaw' slides with its lower horizontal face in contact with the upper face 'ff' of the base plate and with its other faces in contact with the surfaces 'r', 's' and 't' of the strap. The right hand end of this sliding piece is kept in contact with the cam lever by means of the spring, and engagement is obtained by operating the handle of the cam lever through an angle of approximately 90 degrees, thus inducing the vee-jaw to move from right to left across the base plate. Disengagement is effected by returning the cam lever to its original position, and allowing the spring to keep the vee jaw and cam lever in contact by exerting a force on the faces of the stop and buffer.

Do not draw the separate parts as shown, but produce full size, and in first angle projection, the following views of the assembled clamping device:

(a) A sectional elevation, corresponding to the vertical plane EF, shown in the plan views of the base plate and the vee jaw. The view required is that seen when looking in the direction of the arrows.

(b) A complete plan as seen when looking in the direction of the arrow P. This view is to be in correct projection with view (a).

(c) A half end elevation as seen when looking in the direction of the arrow H. The view is to show that part which lies to the left of the vertical central plane.

The vee jaw is to be drawn in the clamped position which corresponds with the dimensions given on the drawing.

No hidden part lines are required in any of the views, and only one of the 12 mm diameter fitted bolts is to be shown.

Insert two dimensions only in each of the views (a) and (b). Draw a title block 100 mm by 60 mm in the bottom right hand corner of your drawing paper and insert the title and scale.

Time allowed – $1\frac{3}{4}$ hours.

Solution (page 162)

The examiner's detailed description of the position and function of each part in the assembly assists considerably.

View (a)

Although this is the first view asked for, it may help if view (b) is drawn first. This will enable the assembly to be built up gradually.

View (b)

The vee jaw is required in the clamping position, i.e., when it is in its most forward position. This will occur when that part of the cam face which is furthest from the cam centre is in contact with the end of the vee jaw. Note that the clearance hole in the strap, where the bolt is omitted, will show two concentric circles – the inner one being one of the circles for the tapped hole in the base.

View (c)

The slope of line 'lm' can be determined by transferring the length of the hidden detail line in view (b).

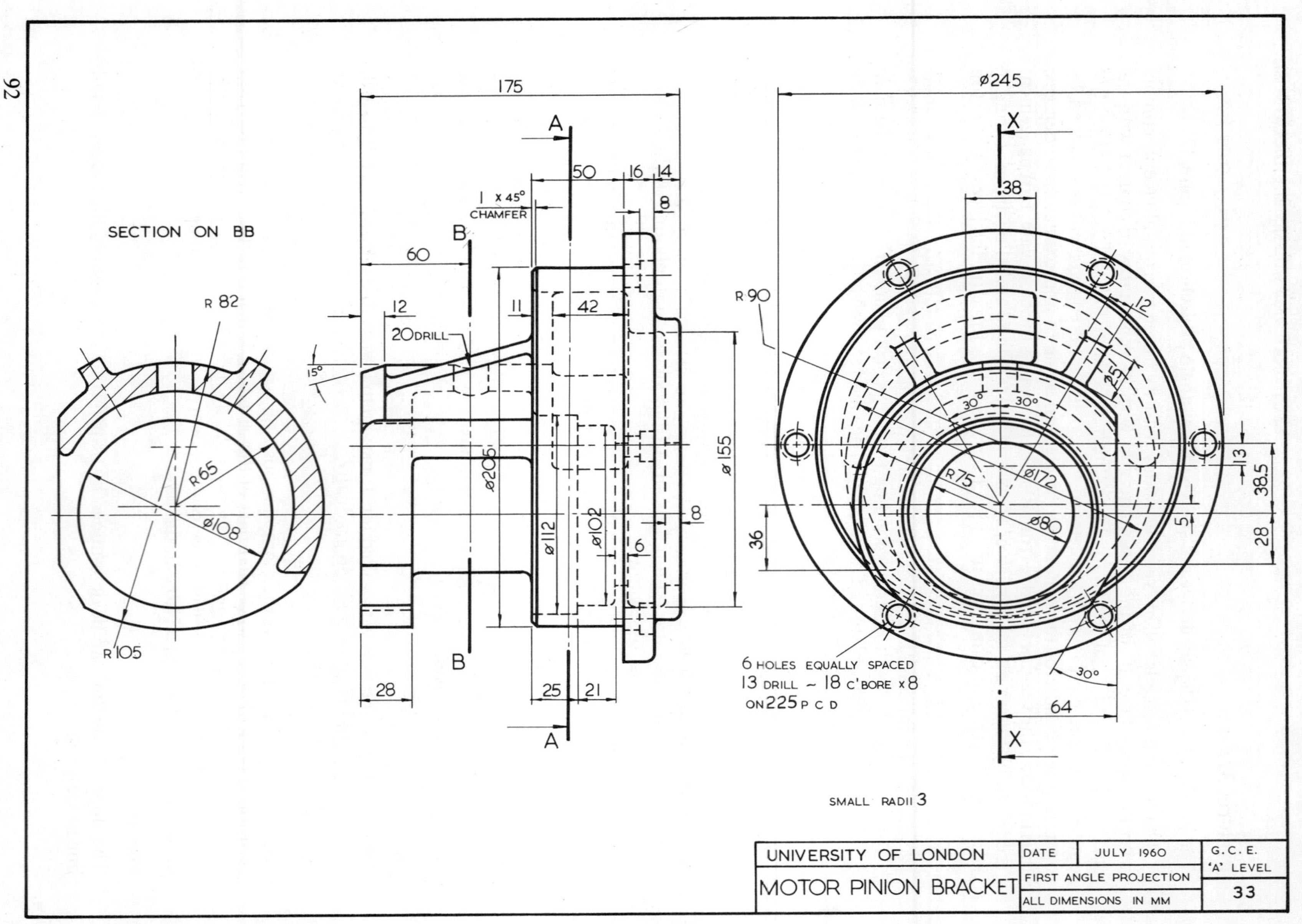

SECTION ON BB
R 82
R65
ø108
R105
175
60
12
20 DRILL
15°
1 x 45° CHAMFER
11
42
50
16
14
8
ø205
ø112
ø102
6
ø155
28
25
21
A
B
ø245
X
38
R 90
12
25
30°
R75
ø172
ø80
13
38.5
28
5
36
64
6 HOLES EQUALLY SPACED
13 DRILL ~ 18 C'BORE x 8
ON 225 P C D
SMALL RADII 3
UNIVERSITY OF LONDON
MOTOR PINION BRACKET
DATE
JULY 1960
FIRST ANGLE PROJECTION
ALL DIMENSIONS IN MM
G.C.E. 'A' LEVEL
33

Problem 33

Figure 33 shows views of a **motor pinion bracket**.

To a scale of full size draw the following views:
(a) A sectional elevation on XX viewed in the direction of the arrows.
(b) A half sectional end elevation to the right of view (a). The plane of the section is to be taken on AA.
(c) A complete plan projected below view (a).
Insert two dimensions only on each view.

N.B. Layout: In order to accommodate the views which are to be drawn in the required positions, set out the axis of the bracket parallel to the shorter edge of your drawing paper, i.e., work with the T square on the longer edge of your drawing board.

Construction lines should be erased and the work neatly lined in. Dotted lines representing hidden parts need not be shown. Sections should be cross hatched in the conventional manner, but pencil shading must not be used. Small radii may be taken as 3 mm, and any missing dimensions should be estimated.

Time allowed – 3 hours.

Solution (page 163)

A careful study is necessary of the considerable amount of hidden detail used to show the complete shape description of the component. The numerous circular shapes of the casting are on four separate centres lying on a vertical centre line and these should be identified in the two given elevations. Although the viewing direction for section BB has not indicated it should be realised that the section has been positioned using first angle projection.

View (a)

This view has a similar profile to the left hand given elevation and since the cutting plane passes through the centre line the sectional view can be built up by changing the hidden detail lines to full lines. The only exception being hidden detail relating to the six counterbored holes.

View (b)

Either half of the component about the vertical centreline may be drawn. All of the four centres are involved in this section and these should be positioned first. It will be necessary to construct lightly one of the two reinforcing ribs. This will enable the ribs to be shown in views (a) and (c).

View (c)

A straightforward projection from view (a). The position of the ribs can be determined from view (b).

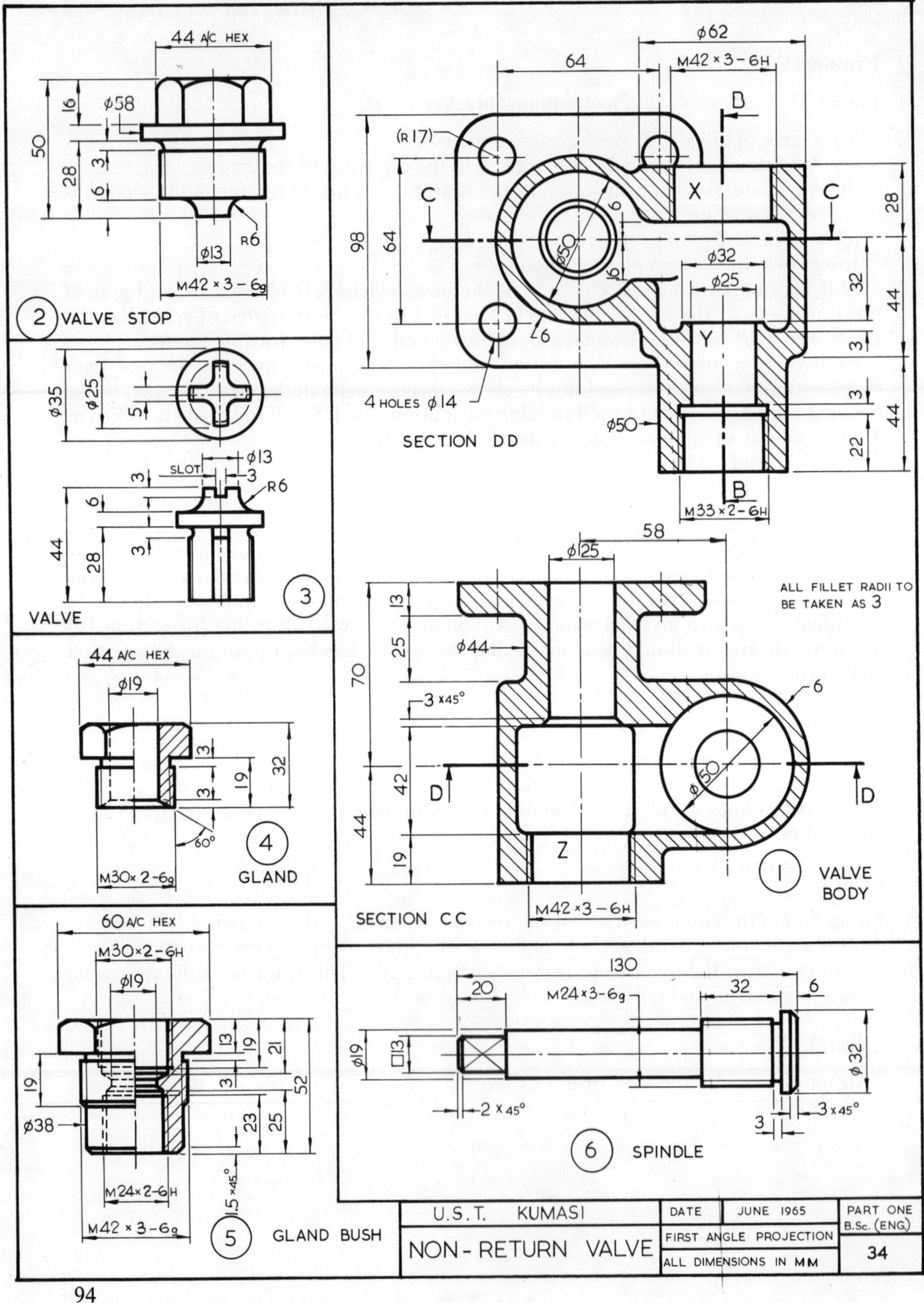

44 A/C HEX
ϕ58
16
50
3
28
6
R6
ϕ13
M42 × 3 – 6g
2 VALVE STOP
ϕ35
ϕ25
5
ϕ13
SLOT
3
R6
6
44
28
3
VALVE
44 A/C HEX
ϕ19
3
19
32
60°
M30× 2-6g
4
GLAND
60 A/C HEX
M30×2-6H
ϕ19
13
19
21
52
23
25
ϕ38
1.5 ×45°
M24×2-6H
M42 × 3-6g
5
GLAND BUSH
ϕ62
64
M42 × 3 - 6H
B
(R 17)
C
X
C
98
64
ϕ50
6
28
ϕ32
ϕ25
32
44
Y
3
4 HOLES ϕ14
ϕ50
22
SECTION DD
B
M33 × 2 - 6H
58
ϕ25
ALL FILLET RADII TO BE TAKEN AS 3
13
25
ϕ44
70
3 x45°
6
42
D
D
ϕ50
44
19
Z
1
VALVE BODY
M42 ×3 – 6H
SECTION CC
130
20
M24 × 3 – 6g
32
6
ϕ19
□13
ϕ32
2 X 45°
3 x45°
3
6
SPINDLE
U.S.T. KUMASI
DATE
JUNE 1965
PART ONE B.Sc. (ENG.)
NON - RETURN VALVE
FIRST ANGLE PROJECTION
ALL DIMENSIONS IN MM
34

Problem 34

Figure 34 shows the details of a **non-return valve**.

Draw full size, using first angle projection, the following views of the assembled valve in the closed position:

(a) A sectional elevation, the plane of the section to be along, and in the direction of, BB.
(b) A sectional plan projected from view A, the plane of the section to be along CC.

Hidden lines need not be shown, but include on the drawing a parts list, title and any other standard data.

Assembly instructions:

The valve (item 3) and valve stop (item 2) are assembled to the body (item 1) at Y and X respectively. The gland bush (item 5) with the spindle (item 6) passing through it is fitted to the body at Z. After the insertion of suitable packing, the gland (item 4) is screwed into the gland bush and adjusted to prevent any fluid leakage via the spindle.

Time allowed – $2\frac{1}{2}$ hours.

Solution (page 164)

View (a)

The section plane BB provides a similar section of the vertical cylindrical portion of the body to that of section plane DD. The portion of the body behind the section plane BB will include outside views of the gland bush, gland and spindle.

View (b)

The lower of the two given sectional views of the body is sectioned on plane CC and can be reproduced, with its correct orientation, for view (b). Items 4, 5, 6 and 7 together form the component parts of a stuffing box.

View (c)

This has been added as an alternative to producing a parts list.

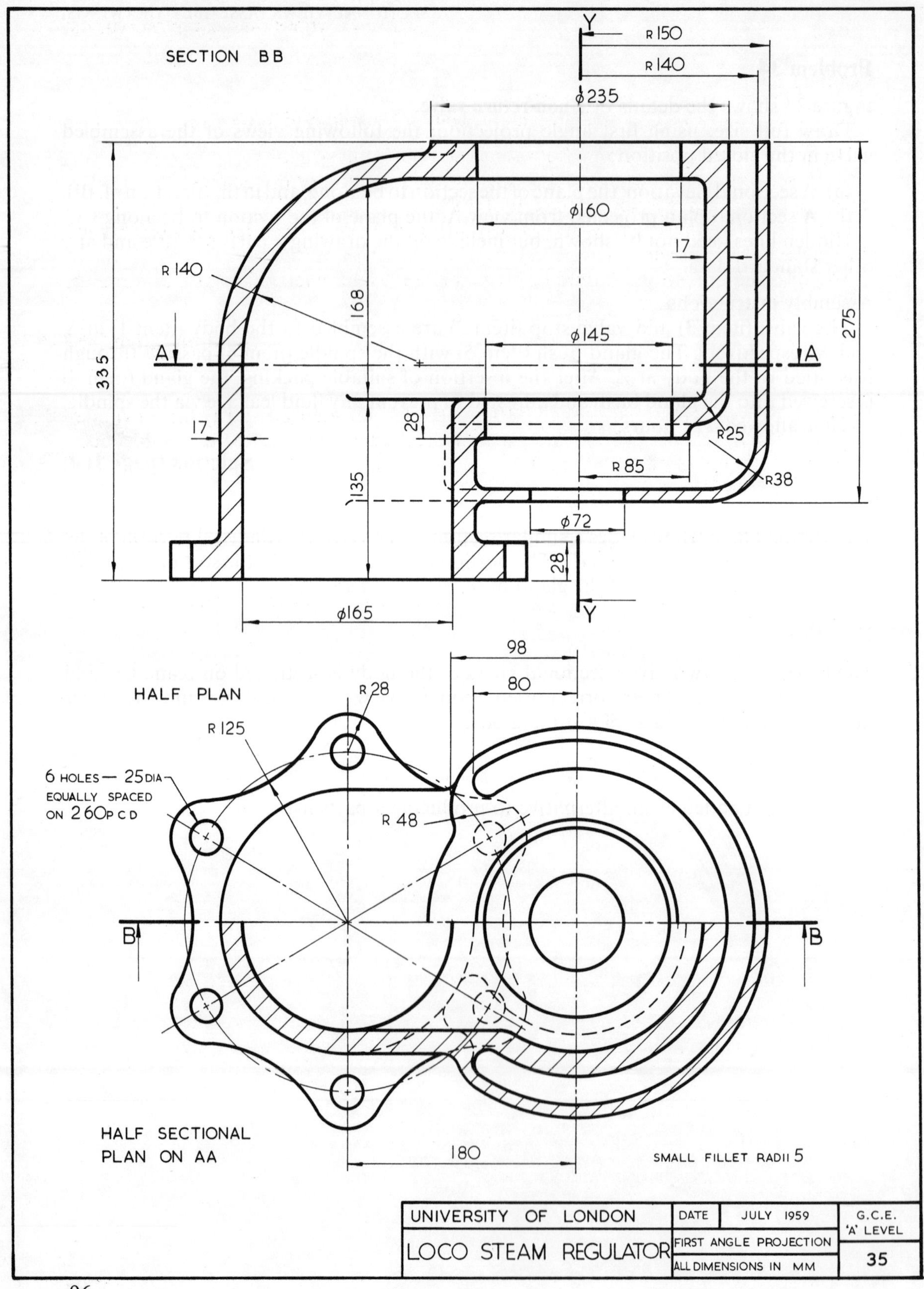
SECTION BB
Y
R 150
R 140
ϕ235
ϕ160
17
R 140
168
275
ϕ145
A
A
335
28
17
R25
R 85
R38
135
ϕ72
28
Y
ϕ165
98
80
HALF PLAN
R 28
R 125
6 HOLES — 25 DIA
EQUALLY SPACED
ON 260 P C D
R 48
B
B
HALF SECTIONAL
PLAN ON AA
180
SMALL FILLET RADII 5
UNIVERSITY OF LONDON
DATE
JULY 1959
G.C.E.
'A' LEVEL
LOCO STEAM REGULATOR
FIRST ANGLE PROJECTION
ALL DIMENSIONS IN MM
35

Problem 35

Figure 35 shows views of a **steam regulator body** for a locomotive.

Do not draw the views given, but to a scale of half full size draw:
(a) A sectional elevation on YY as seen in the direction of the arrows.
(b) An elevation projected to the left of (a).
The work should be neatly and clearly lined in. Dotted lines representing hidden parts need not be shown.
Any missing dimensions should be estimated and small fillets taken as 4 mm radius.
Time allowed – 3 hours.

Solution (page 165)

View (a)

Section YY reveals a thin outer shell – 10 mm thick – shrouding a sturdy inner cylinder – 17 mm thick. Details of their merger and their attachment to the elbow has been shown in the half sectional plan on AA.

View (b)

Either angle of projection may be used since the regulator is symmetrical, and either of its sides presents a similar test.
The curve of intersection can be resolved into the result of the intersection of two parallel cylinders.

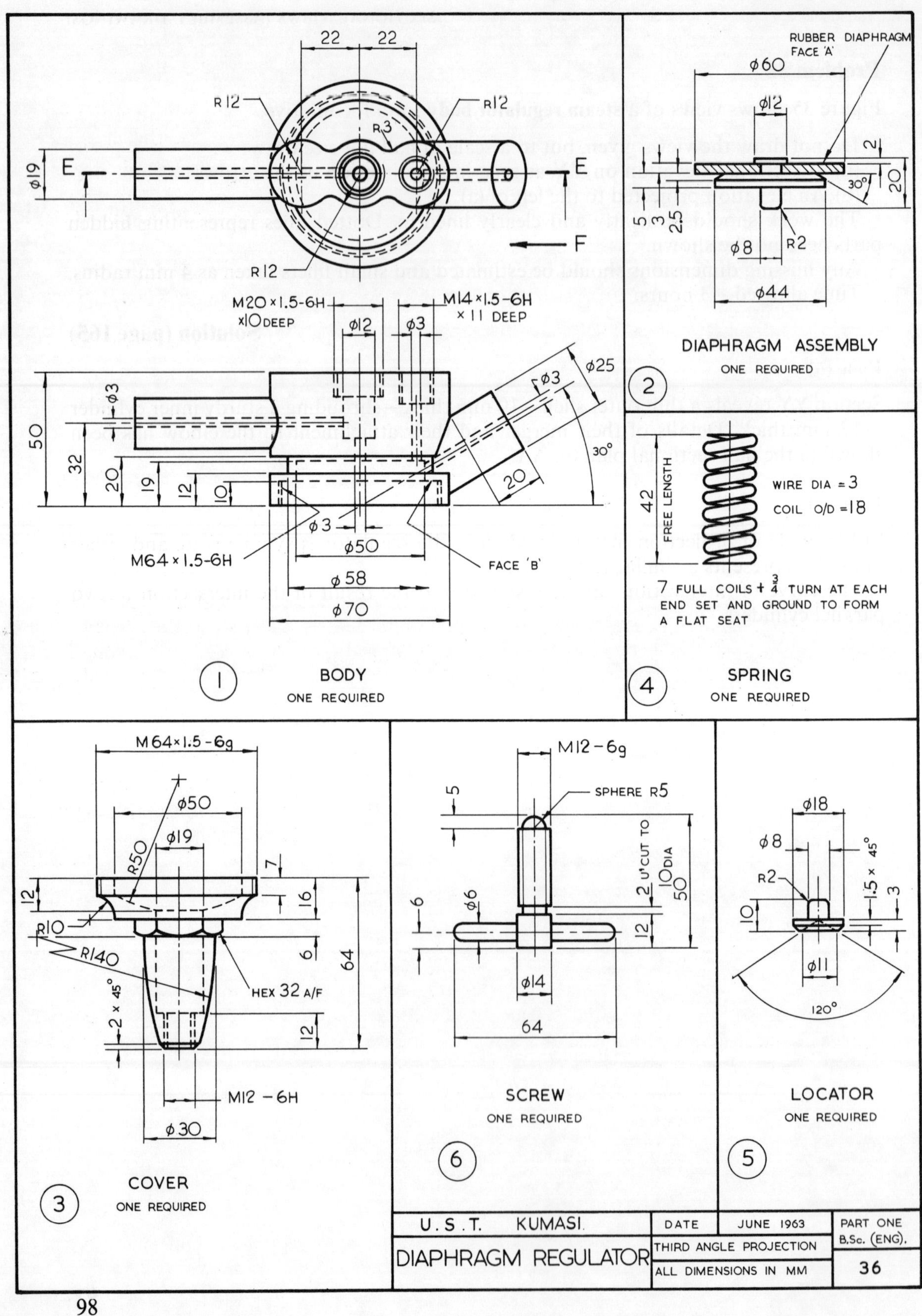

22 22
R12
R12
R3
φ19
E
E
F
R12
M20×1.5-6H ×10 DEEP
M14×1.5-6H ×11 DEEP
φ12
φ3
φ25
φ3
φ3
30°
50
32
20
19
12
10
20
φ3
φ50
M64×1.5-6H
FACE 'B'
φ58
φ70
1
BODY
ONE REQUIRED
RUBBER DIAPHRAGM FACE 'A'
φ60
φ12
2
20
30°
5
3
25
φ8
R2
φ44
DIAPHRAGM ASSEMBLY
ONE REQUIRED
2
42
FREE LENGTH
WIRE DIA = 3
COIL O/D = 18
7 FULL COILS + 3/4 TURN AT EACH END SET AND GROUND TO FORM A FLAT SEAT
4
SPRING
ONE REQUIRED
M64×1.5-6g
φ50
φ19
R50
12
7
16
R10
R140
HEX 32 A/F
2 × 45°
6
64
12
M12 -6H
φ30
COVER
ONE REQUIRED
3
M12-6g
SPHERE R5
5
2 U'CUT TO 10 DIA
50
6
φ6
12
φ14
64
SCREW
ONE REQUIRED
6
φ18
φ8
1.5 × 45°
R2
10
3
φ11
120°
LOCATOR
ONE REQUIRED
5
U.S.T. KUMASI
DATE
JUNE 1963
PART ONE B.Sc. (ENG).
DIAPHRAGM REGULATOR
THIRD ANGLE PROJECTION
ALL DIMENSIONS IN MM
36

Problem 36

Figure 36 shows details of part of a **diaphragm regulator**.

The diaphragm assembly (item 2) is inserted into the body (item 1) with face A of item 2 against face B of item 1. The spring (item 4) is then fitted with the 8 mm dia peg in the spring. The cover (item 3) is then screwed into the body and the screw (item 6) is fitted to the cover, its ball end seating in the 120° conical seat of the locator and adjusted so that the spring is compressed to a length of 35 mm.

Do not copy the given views but draw full size the following views, using third angle projection, of the diaphragm regulator with the parts assembled in their correct relative positions:

(a) Sectional elevation, the plane of the section and the direction of the required view being shown at EE.

(b) Elevation looking in the direction of the arrow F.

Hidden part lines are not required in any of the views.

Time allowed – 2 hours.

Solution (page 166)

View (a)

The cutting plane passes through the centre of the body and as a consequence the sectional view of item 1 can be determined by converting the given hidden detail lines to full lines. The given views of item 2 can now be reproduced in correct position in item 1. Converting the given hidden detail lines of item 3 to full lines will give its sectional view. It will of course be necessary to construct lightly a hexagon of 32 mm A/F in order that the across corners size can be determined. Note that the screw has not been sectioned, in accordance with BS 308.

View (b)

Most components visible in this view appear as diameters, and light construction lines will be necessary to determine the ellipse representing the end face of the sloping branch pipe.

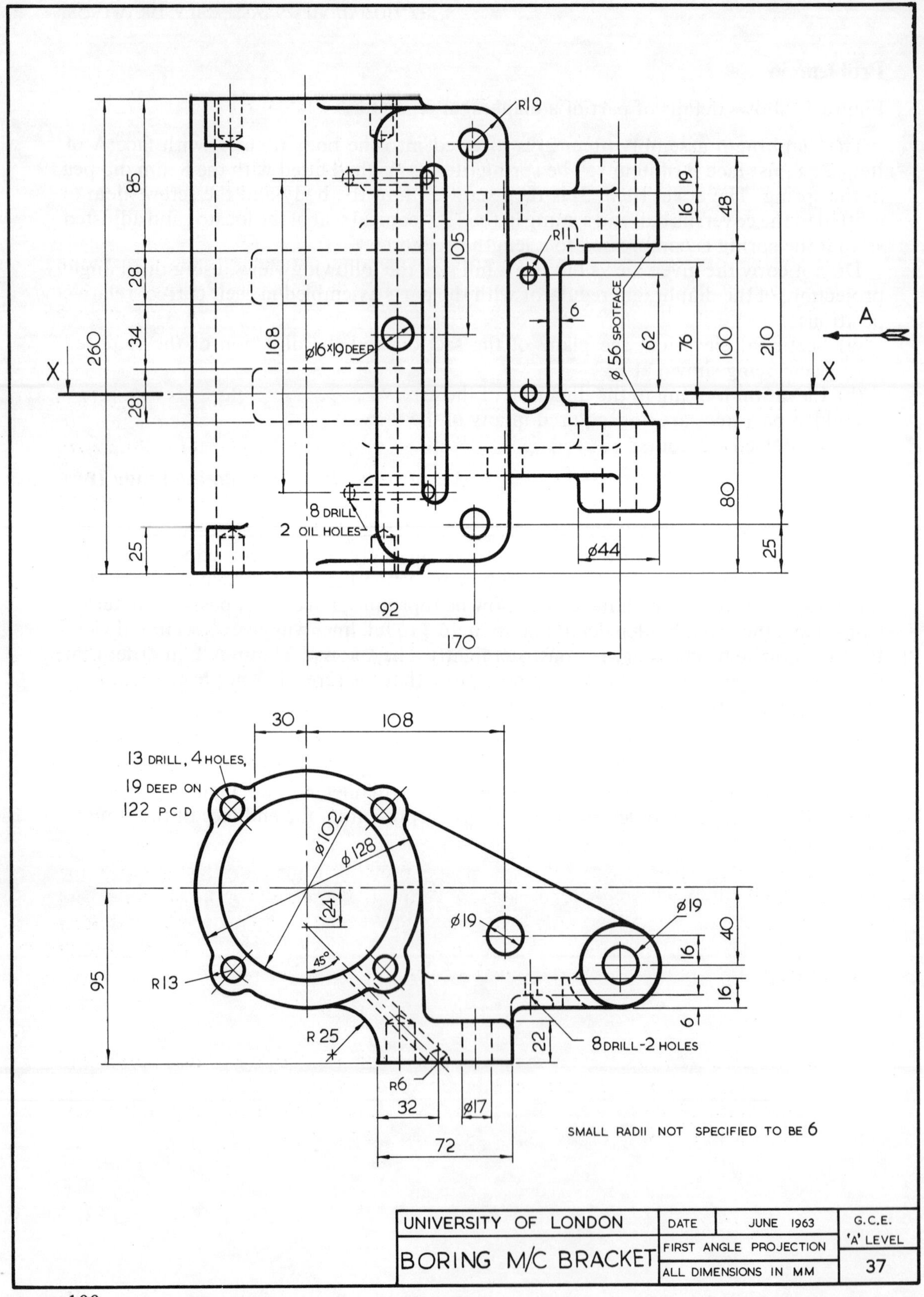
R19
85
19
48
16
105
R11
28
6
ø 56 SPOTFACE
62
76
100
210
A
34
168
260
ø16 x 19 DEEP
X
X
28
8 DRILL
2 OIL HOLES
80
25
ø44
25
92
170
30
108
13 DRILL, 4 HOLES,
19 DEEP ON
122 P C D
ø102
ø128
(24)
ø19
ø19
40
16
16
6
95
R13
45°
R 25
22
8 DRILL-2 HOLES
R6
32
ø17
72
SMALL RADII NOT SPECIFIED TO BE 6
UNIVERSITY OF LONDON
BORING M/C BRACKET
DATE
JUNE 1963
FIRST ANGLE PROJECTION
ALL DIMENSIONS IN MM
G.C.E.
'A' LEVEL
37

Problem 37

Two views of a **bracket for a boring machine** are given in Figure 37.

Draw to a scale of $\frac{3}{4}$ full size the following views:
(a) An elevation as seen in the direction of the arrow A.
(b) A side elevation projected from (a) showing the reverse side of the bracket.
(c) A sectional plan on XX projected below (b).
Insert six important dimensions.

Omit broken lines indicating hidden parts. Erase all construction lines, and line in the work neatly. Sections should be cross hatched in the conventional manner. Any missing dimensions should be estimated, and small radii drawn with bow compasses set at 3 mm.

Time allowed – 3 hours.

Solution (page 167)

The student is advised to sketch small views of the component using simple blocks to represent each view, in order to record his interpretation of the orientation of the views asked for. These should then be studied to check that they conform to the viewing directions required and that they satisfy either first or third angle projection.

View (b)

It is probably better to start with this view and project view (a) from it.

View (b) is the reverse side to the upper of the two given elevations and because of the required projection sequence, will be of the opposite hand. A careful study of the given hidden detail will reveal that the main 128 mm diameter tube has two slots 28 mm wide and these will appear in full outline in view (b), as will the two 16 mm thick webs. It will be necessary for view (c) to be drawn next in order that the protruding lugs of 13 mm radius can be projected into view (b).

View (c)

The cutting plane passes through one of the previously mentioned slots with the resultant reduction in cross hatched surface. Note that the 56 mm dia spotface tool removes metal over a small arc only but sufficient to enable adequate clearance for any mating pivot piece fitted between the two machined faces 100 mm apart. Because of the position of the cutting plane the fourth protruding lug is not visible.

View (a)

The student is advised to draw the nearest features first, i.e., the two 44 mm bosses and their supporting webs. The 128 mm dia tube can then be projected from view (b) including the appropriate cutaways resulting from the 16 mm wide slots. Complete the view by projecting a side view of the 72 mm × 248 mm base plate.

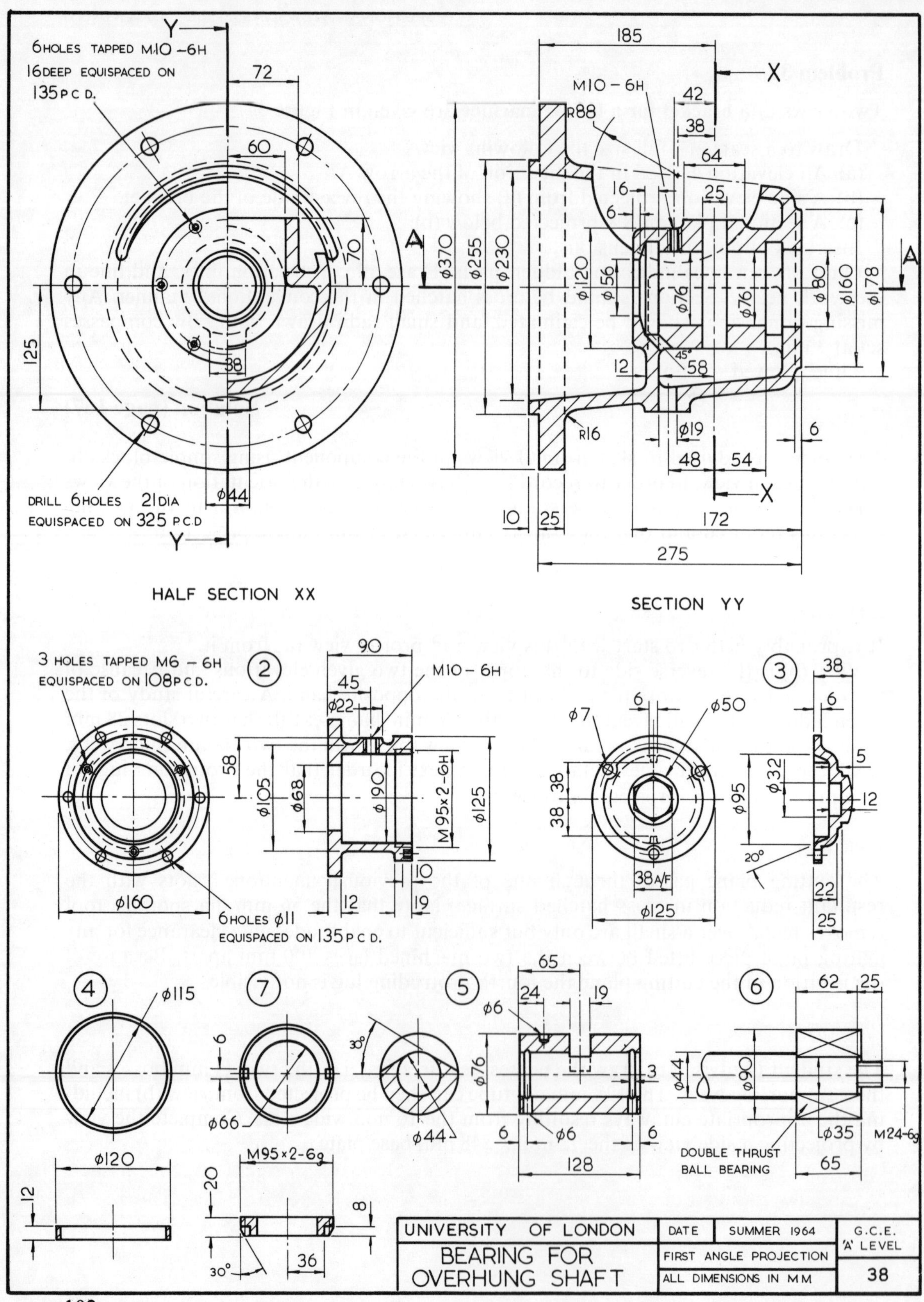
6 HOLES TAPPED M10 - 6H
16 DEEP EQUISPACED ON
135 P C D.
DRILL 6 HOLES 21 DIA
EQUISPACED ON 325 P C D
HALF SECTION XX
SECTION YY
M10 - 6H
R88
R16
3 HOLES TAPPED M6 - 6H
EQUISPACED ON 108 P C D.
M10 - 6H
M95 x 2 - 6H
6 HOLES ϕ11
EQUISPACED ON 135 P C D.
38 A/F
M95 x 2 - 6g
DOUBLE THRUST
BALL BEARING
M24-6g
UNIVERSITY OF LONDON
BEARING FOR
OVERHUNG SHAFT
DATE
SUMMER 1964
FIRST ANGLE PROJECTION
ALL DIMENSIONS IN MM
G.C.E.
'A' LEVEL
38

Problem 38

The drawings on Figure 38 show the detail components of a combined ring-oiled bearing and ball thrust **bearing for an overhung shaft**.

Assembly

The shaft is supported in the bronze bush (5) which is fixed in position in the main casing (1) by means of a M10–6H grub screw. This bush is lubricated by the ring (4) which rides on the shaft and dips into the oil reservoir in the casing. Axial load on the shaft is taken on a double thrust ball bearing (6) fitted on the portion of the shaft reduced to 35 mm dia and secured with two 24 mm locknuts. This bearing is enclosed in the housing (2) which is attached to the smaller end of the main casing with six M10 studs and nuts. The screwed ring (7) retains the thrust bearing in position and is tightened by means of the key lugs on the cover (3). The cover is secured to the housing with three M6–6g cheesehead screws.

Using a scale of half full size draw the following views of the complete arrangement:

(a) An outside view corresponding to the section YY.

(b) The plan view, half of which should be in section taken on the centre line AA.

The component parts should be shown in their respective working positions. All construction lines should be erased except those necessary to produce curves of inter-penetration. Sections should be cross hatched in the conventional manner, but pencil shading or colour washing must not be used. Omit broken lines indicating hidden parts. Dimensions are not required. View (a) need not be lined in. Small unspecified radii may be taken as 4 mm.

Time allowed – 3 hours.

Solution (page 168)

Figure 38A illustrates the arrangement of the shaft, bronze bush and oil ring. The shaft causes the oil ring to rotate and carry with it a thin film of oil which eventually returns to the reservoir via the annulus and 6 mm dia hole at each end of the bronze bush.

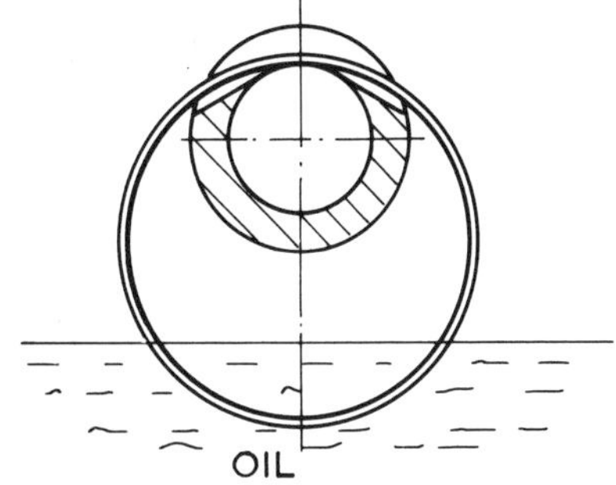

Figure 38A

View (a)

Only six of the twelve component types which make up the assembly are seen in this outside view. These are items 1, 2 and 3 and their securing screws, studs and nuts. In order that the screw and stud positions may be determined for the two required views it will be necessary to construct lightly, adjacent to view (a), the two pitch circle diameters and mark out the screw and stud positions.

View (b)

The construction necessary for the shape of the large cut away in the half that is drawn as an outside view requires that vertical sections are taken through the conical shell of the main casing and that a scrap end elevation be projected from view (a).

A scrap elevation is also necessary for the half that is drawn in section on AA to determine where the oil ring is cut by the section plane.

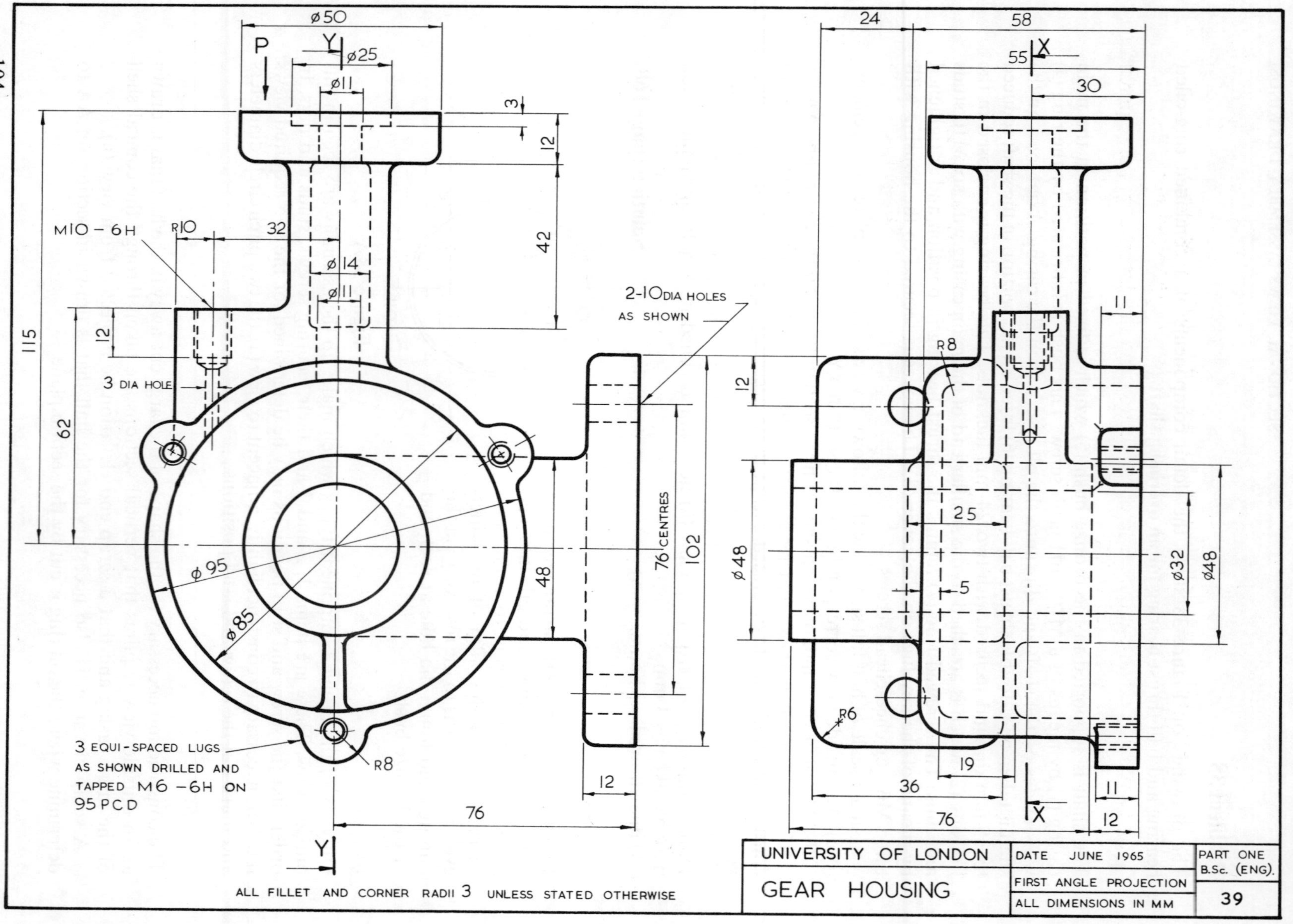

P
Y
ø50
ø25
ø11
ø14
ø11
M10 - 6H
R10
32
3 DIA HOLE
2-10DIA HOLES
AS SHOWN
76 CENTRES
ø95
ø85
R8
3 EQUI-SPACED LUGS
AS SHOWN DRILLED AND
TAPPED M6 -6H ON
95 PCD
X
R6
ø48
ø32
ALL FILLET AND CORNER RADII 3 UNLESS STATED OTHERWISE
UNIVERSITY OF LONDON
GEAR HOUSING
DATE JUNE 1965
FIRST ANGLE PROJECTION
ALL DIMENSIONS IN MM
PART ONE
B.Sc. (ENG).
39

Problem 39

Figure 39 shows two orthographic views of a **cast aluminium gear housing**.

Do not copy these views as shown but draw the following views full size:

(a) A sectional elevation, the plane of the section and the direction of the required view being as indicated at XX.

(b) A sectional elevation, the plane of the section and the direction of the required view being indicated at YY.

(c) A complete plan, projected from view (a), as seen in the direction of arrow P.

Hidden part lines are to be shown in view (c) only. Insert on the drawings six important dimensions, scale and method of projection used.

The views may be drawn in either (but not both) first or third angle projection.

Time allowed – 2 hours.

Solution (page 169)

View (a)

The viewing direction of Section XX is similar to that of the left hand given elevation and its profile can be copied. The only features in front of the cutting plane, and consequently not seen, are the three lugs.

View (b)

The required section has a similar viewing direction to that of the right hand given elevation, and its profile also may be copied. The one feature lost in front of the cutting plane is the portion containing the M10 tapped hole.

View (c)

Build up the view by drawing the nearest shapes first, i.e., the 50 mm dia flange, followed by the main circular body and its projecting tube 48 mm dia. The right hand extension piece needs careful study to visualise shape description and it should be noted that the thickness, i.e., 25 mm, is given on the right hand given view.

It will be necessary to construct lightly the three lugs in view (a) in order that their plan view may be drawn in view (c).

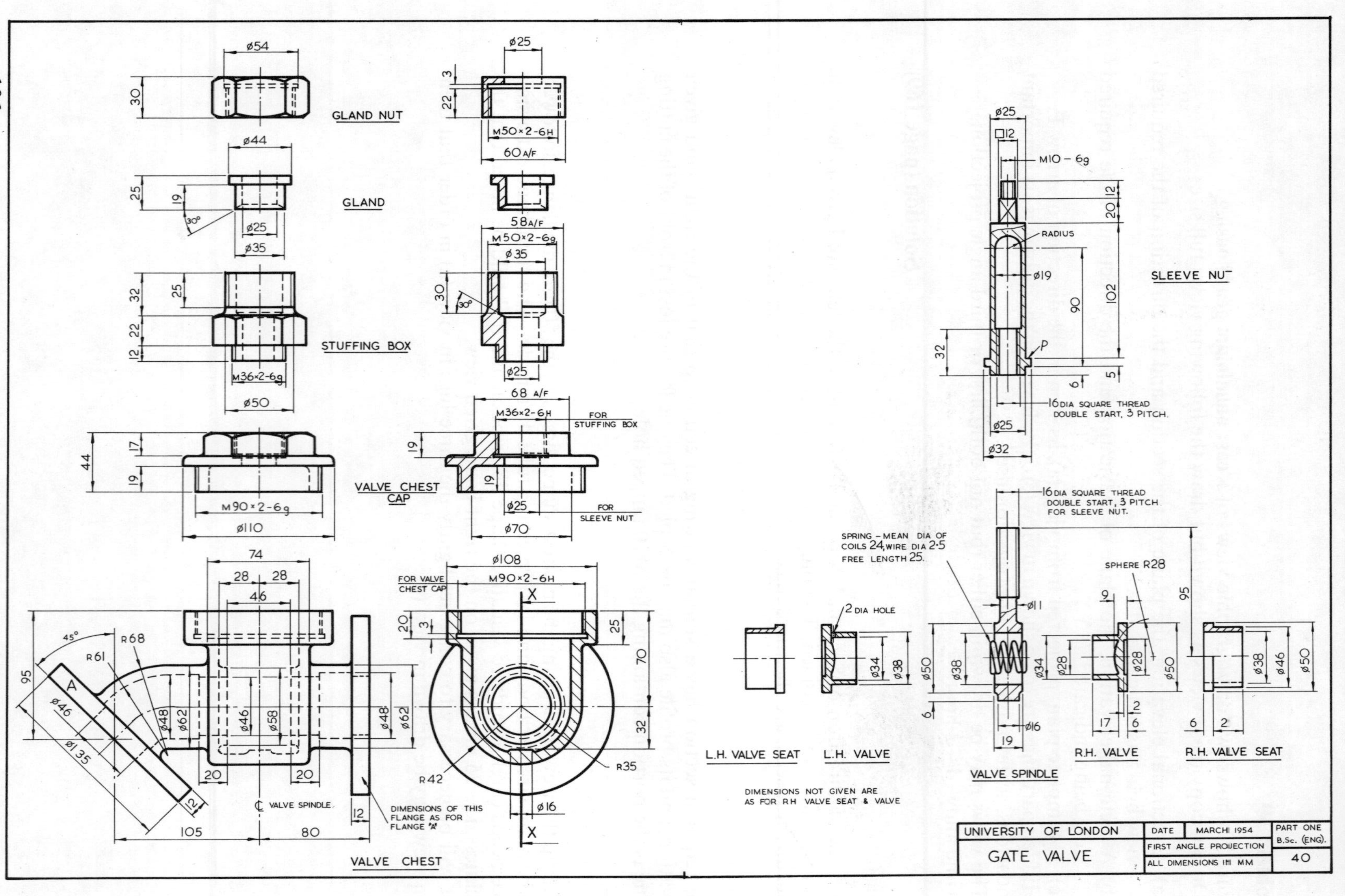
GLAND NUT
GLAND
STUFFING BOX
VALVE CHEST CAP
VALVE CHEST
SLEEVE NUT
VALVE SPINDLE
L.H. VALVE SEAT
L.H. VALVE
R.H. VALVE
R.H. VALVE SEAT
M50×2-6H
60 A/F
58 A/F
M50×2-6g
M36×2-6g
68 A/F
M36×2-6H
FOR STUFFING BOX
FOR SLEEVE NUT
M90×2-6g
M90×2-6H
FOR VALVE CHEST CAP
DIMENSIONS OF THIS FLANGE AS FOR FLANGE 'A'
℄ VALVE SPINDLE
M10 – 6g
RADIUS
16 DIA SQUARE THREAD DOUBLE START, 3 PITCH.
16 DIA SQUARE THREAD DOUBLE START, 3 PITCH FOR SLEEVE NUT.
SPRING – MEAN DIA OF COILS 24, WIRE DIA 2·5 FREE LENGTH 25.
SPHERE R28
2 DIA HOLE
DIMENSIONS NOT GIVEN ARE AS FOR RH VALVE SEAT & VALVE
UNIVERSITY OF LONDON
GATE VALVE
DATE
MARCH 1954
FIRST ANGLE PROJECTION
ALL DIMENSIONS IN MM
PART ONE B.Sc. (ENG).
40

Problem 40

Details of the component parts of a **gate valve** are shown in Figure 40.

Draw full size, using first angle projection, the following views of the assembled valve:

(a) A sectional elevation corresponding to the section plane XX shown on the given views of the valve chest.

(b) An outside elevation projected to the right hand side of view (a).

Assembly instructions

The components have been drawn in a way that illustrates their relative positions. It should be noted that the sleeve nut is retained in the valve chest cap by surface 'p' of the collar on the sleeve nut and the underside of the stuffing box.

Time allowed – 2 hours.

Solution (page 170)

View (a)

The question paper illustrates all parts in an 'exploded' position, and this helps in determining the assembled position for each component. The section plane cuts through the centre of symmetrical components, and their sectional views can be achieved by changing the given hidden detail lines to full lines. The top of the sleeve nut has been shown in full to conform with recommendations made in BS 308 regarding the sectioning of such parts as bolts, screws, shafts, etc.

View (b)

This view has been chosen for illustrating the hexagonal parts in the across corner position – the reason for this choice being that it tends to give a more balanced appearance and results in thinner wall sections to be cross hatched in view (a), i.e., across flats.

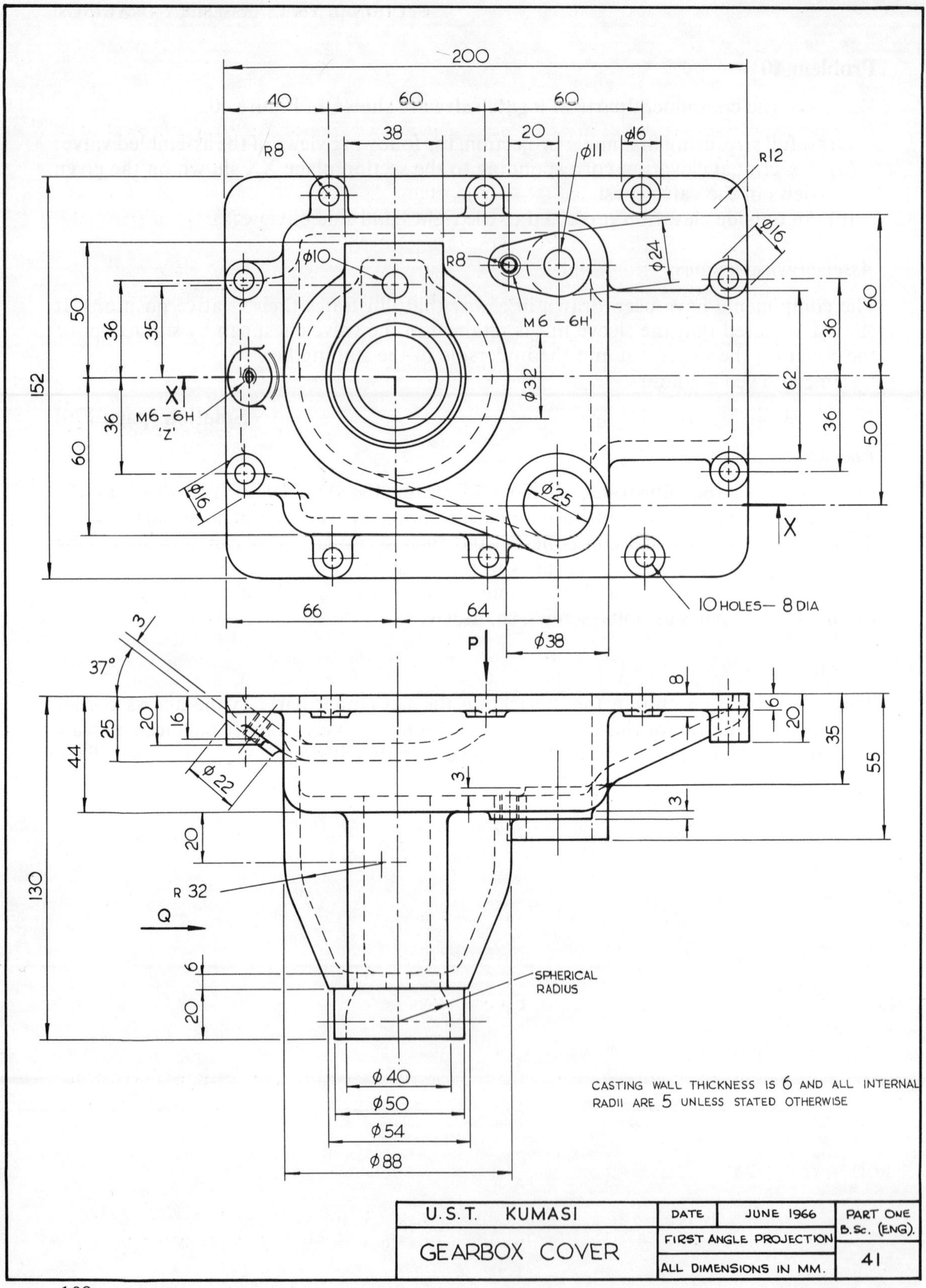
200
40
60
60
38
20
φ16
R8
φ11
R12
φ10
R8
φ24
φ16
M6 – 6H
50
36
35
60
36
62
152
X
M6 – 6H
'Z'
φ32
36
60
36
50
φ16
φ25
X
10 HOLES— 8 DIA
66
64
φ38
3
37°
P
8
6
20
35
55
20
16
25
44
φ22
3
3
20
130
R 32
Q
6
SPHERICAL
RADIUS
20
φ40
φ50
φ54
φ88
CASTING WALL THICKNESS IS 6 AND ALL INTERNAL
RADII ARE 5 UNLESS STATED OTHERWISE
U.S.T. KUMASI
DATE
JUNE 1966
PART ONE
B.Sc. (ENG).
GEARBOX COVER
FIRST ANGLE PROJECTION
ALL DIMENSIONS IN MM.
41

Problem 41

Two views are shown in first angle projection in Figure 41.

Do not copy these views but draw full size, using either first or third angle projection, the following views:

(a) A sectional view as indicated by section plane XX.
(b) A half outside view as seen in the direction of arrow Q, projected from view (a). The half view required is to the right of the centre line.
(c) A bottom view as seen in the direction of arrow P, projected from view (a).

In view (a) only, show a M6 × 32 mm long stud and a full hexagon nut at the hole marked Z.

Hidden detail is not required in any view.

Print the title, scale and method of projection used.

Time allowed – 2 hours.

Solution (page 171)

View (a)

At one position the offset cutting plane XX passes through the casting at an acute angle. The shape description of the casting at this position, and the construction necessary for determining Section XX, are shown in Figure 41A.

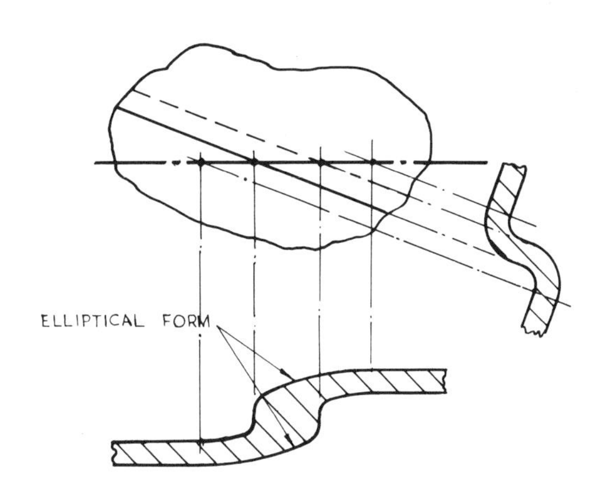

Figure 41A

View (b)

This view requires the construction of a number of curves of intersection. These can be resolved into the intersection of simple shapes, i.e., cylinder/cylinder and cylinder/prism.

View (c)

An inverted plan viewed in the opposite direction to the upper of the two given views. The given hidden detail lines become full lines and the given full lines are omitted.

Problem 42

Figure 42 is a detail drawing of a **self-centring chuck for a lathe**.

Three jaw blocks F are mounted in position in the slotted grooves of the face plate B and equidistant from the centre O.

A scroll plate D is fitted with the face 'a' of D making contact with face 'b' of B. The rack teeth of blocks F are engaged by the scroll D.

Three pinions C are then mounted in position with the teeth engaging those on the back of D. The cover plate A is attached with the faces 'm' and 'n' in contact. It is secured by means of nine 6 mm diameter screws H.

The flange E is fastened by means of three 10 mm diameter screws G to the back of A with the surfaces 'p' and 'q' making contact. The flange and chuck are secured directly to the end of the lathe spindle by means of the 50 mm diameter screw thread in E. When assembled in this way it is possible to rotate any one of the three pinions about its axis. This causes D to rotate, the blocks F move in a radial direction and the work is held as in a vice.

(a) Sketch freehand and in good proportion the solution asked for in part (b). Use the squared paper supplied for the purpose and finally secure it to the left hand top corner of the drawing paper.

(b) Showing the parts in their respective working positions draw two views:
  (i) The side view of the assembled chuck and flange, i.e., the view corresponding with that seen in the direction of the arrow P.
  (ii) The view seen when looking from the right hand side of (i).

In view (i) above the centre line show the parts A, B, D and E in section and other parts in outside view. Below the centre line show the outside view of the whole assembly. Also in (i) let the work W be held by the chuck and shown in position as a phantom view using chain dotted lines. Show the face 'c' of blocks F making contact with the internal diameter of W and the face 'e' of W touching the face 'd' of F.

Omit the work W in view (ii).

Scale: full size.

The drawing should be neatly and clearly lined in, construction lines being erased. Dotted lines may be omitted.

Insert twelve of the more important dimensions. Print the title of the drawing: self-centring chuck for lathe, height of printing to be 6 mm.

Time allowed – 2 hours.

Solution (page 172)

Part (a) of the question paper has not been included in the solution on page 172 but it is recommended that the student attempts this very useful exercise.

View (i)

The detailed and informative assembly instructions given in the question paper should enable the student to gradually build up a mental picture of the position of each of the twenty two components which make up the lathe chuck assembly. The student should not be put off by the fact that a number of the components, namely parts A and B, contain a great deal of information. A small percentage only of this information is actually drawn in the solution and once the position for each component has been decided upon a start should be made on view (i). It is recommended that the outside elevations of parts A, B and E below their horizontal centre line are drawn first, and their sectional views then built up above the centre line.

View (ii)

This view is straightforward and, it is felt, needs no comment.

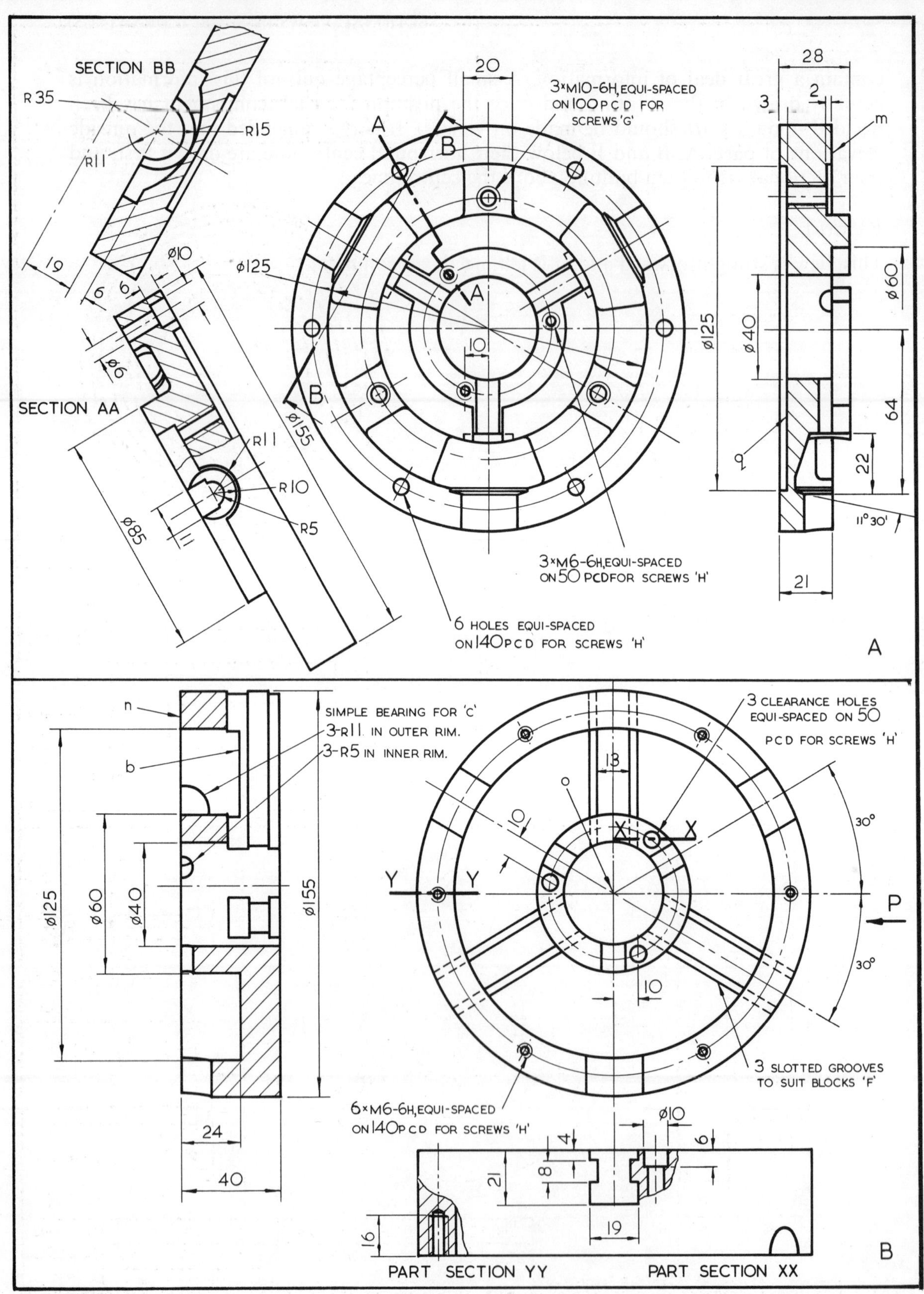
SECTION BB
R35
R15
R11
19
6
6
ø10
ø6
SECTION AA
R11
R10
R5
ø85
11
ø155
ø125
20
3×M10-6H, EQUI-SPACED
ON 100 P C D FOR
SCREWS 'G'
A
B
A
B
10
3×M6-6H, EQUI-SPACED
ON 50 PCD FOR SCREWS 'H'
6 HOLES EQUI-SPACED
ON 140 P C D FOR SCREWS 'H'
28
2
3
m
ø60
ø40
ø125
64
22
q
11°30'
21
A
n
b
SIMPLE BEARING FOR 'C'
3-R11 IN OUTER RIM.
3-R5 IN INNER RIM.
ø125
ø60
ø40
ø155
24
40
3 CLEARANCE HOLES
EQUI-SPACED ON 50
P C D FOR SCREWS 'H'
13
o
10
X
X
Y
Y
P
30°
30°
10
3 SLOTTED GROOVES
TO SUIT BLOCKS 'F'
6×M6-6H, EQUI-SPACED
ON 140 P C D FOR SCREWS 'H'
ø10
4
6
8
21
19
16
PART SECTION YY
PART SECTION XX
B

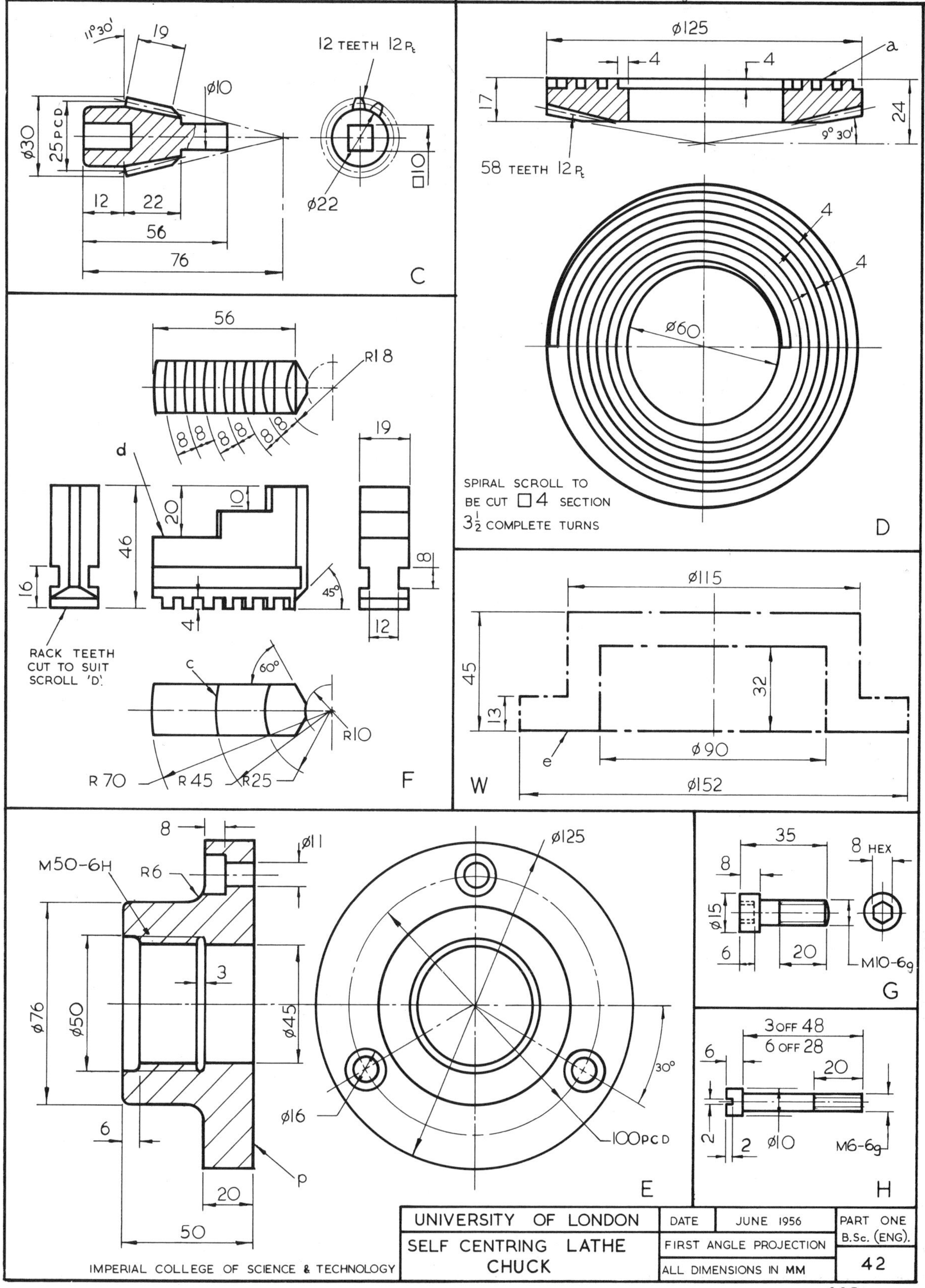
11°30'
19
12 TEETH 12 Pt
ø10
ø30
25 PCD
□10
ø22
12
22
56
76
C
ø125
4
4
a
17
24
9° 30'
58 TEETH 12 Pt
4
4
ø60
SPIRAL SCROLL TO
BE CUT □4 SECTION
3½ COMPLETE TURNS
D
56
R18
8 8 8 8 8 8
19
d
20
10
46
16
8
45°
4
12
RACK TEETH
CUT TO SUIT
SCROLL 'D'
c
60°
R10
R70
R45
R25
F
ø115
45
32
13
e
ø90
ø152
W
8
ø11
M50-6H
R6
ø125
ø76
ø50
3
ø45
30°
ø16
100 PCD
6
p
20
50
E
35
8 HEX
8
ø15
6
20
M10-6g
G
3 OFF 48
6 OFF 28
6
20
2
2
ø10
M6-6g
H
UNIVERSITY OF LONDON
DATE
JUNE 1956
PART ONE
B.Sc. (ENG).
SELF CENTRING LATHE
CHUCK
FIRST ANGLE PROJECTION
ALL DIMENSIONS IN MM
42
IMPERIAL COLLEGE OF SCIENCE & TECHNOLOGY

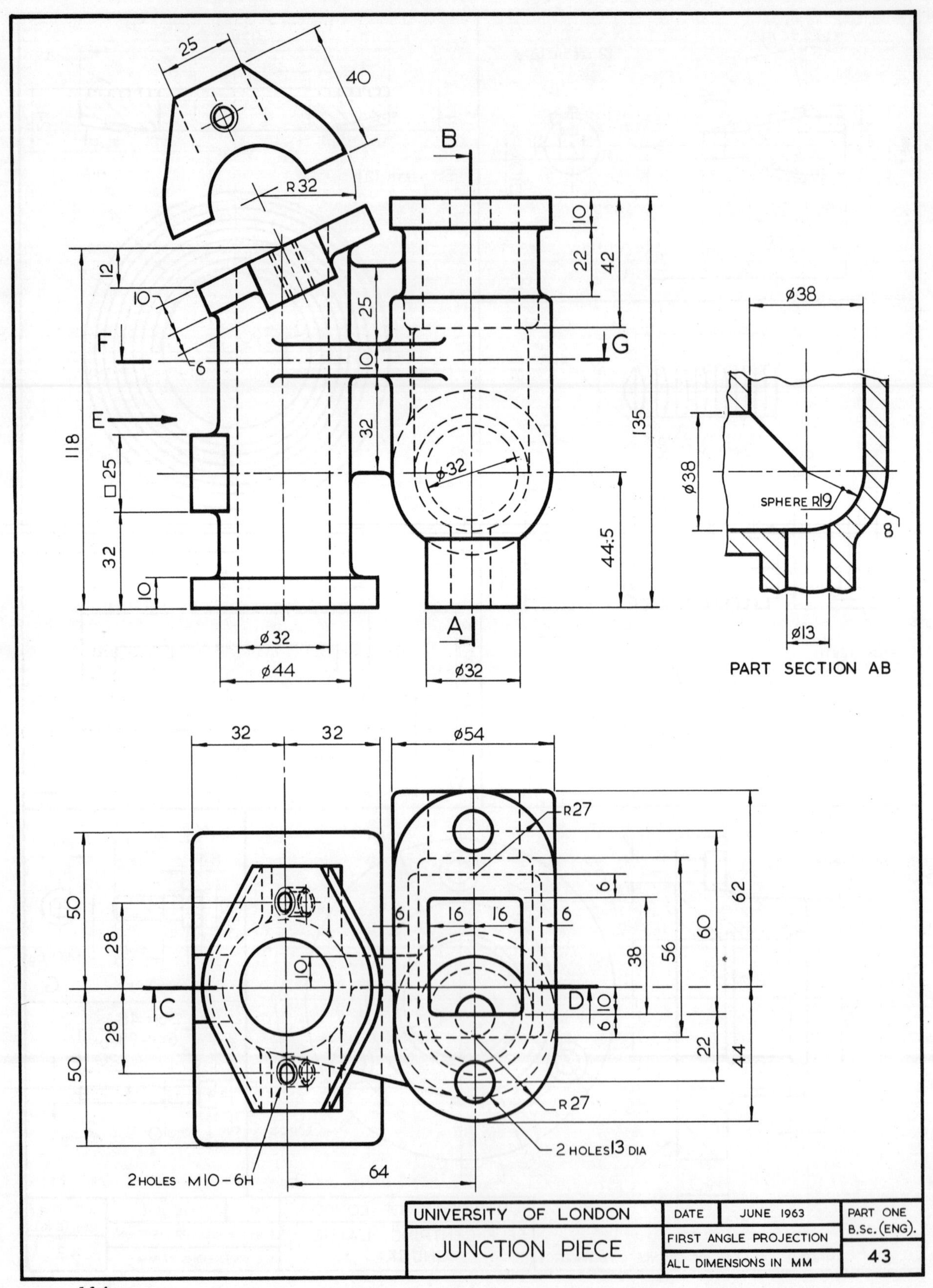

25
40
R 32
B
A
E
F
G
C
D
12
10
6
118
□ 25
32
10
25
10
32
22
42
135
44·5
ø32
ø32
ø44
ø32
ø38
ø38
SPHERE R19
8
ø13
PART SECTION AB
32
32
ø54
R27
50
28
28
50
6
16
16
6
10
6
10
6
38
56
60
62
22
44
R27
2 HOLES 13 DIA
2 HOLES M10-6H
64
UNIVERSITY OF LONDON
JUNCTION PIECE
DATE
JUNE 1963
FIRST ANGLE PROJECTION
ALL DIMENSIONS IN MM
PART ONE
B.Sc. (ENG).
43

Problem 43

Orthographic views of a **junction piece** are shown in Figure 43.

Do not copy the views as shown but draw full size the following:

(a) A sectional elevation, the plane of the section and the direction of the required view being indicated at CD.
(b) An outside elevation as seen when looking in the direction of the arrow E.
(c) A sectional plan, the plane of the section and the direction of the required view being indicated at FG.

Insert on the drawing six important dimensions and in the lower right hand corner of the paper draw a title block 100 mm by 60 mm and insert relevant data.

Hidden part lines are to be shown on view (c) only. The views may be drawn in either first or third angle projection and the method of projection used must be stated in the title block.

Time allowed – 2 hours.

Solution (page 173)

The junction piece comprises two basic units: a 44 mm diameter tube surmounted by an inclined faceplate and a 54 mm diameter elbow surmounted by a rectangular tube and oval flange. These two units are connected by two webs perpendicular to each other.

View (a)

The cutting plane passes through the centre of the tubular section of the left hand unit along the front face of the vertical connecting web and through the elbow but at a distance of 10 mm from the centre of its vertical leg.

The offset cutting plane results in the 'necking' of the vertical leg of the elbow, and the projected width should be found to be similar to the bore of the rectangular tube, i.e., 32 mm. Note that it is necessary to cross hatch one of the connecting webs, and that the cutting plane does pass through the centre of the elbow supporting boss.

View (b)

In order that the inclined faceplate may be drawn in view (b), it will be necessary to construct an auxiliary view of the faceplate similar to that given on the question paper.

The pictorial view and the hidden part lines in view (b) have been included as an aid to visualisation.

View (c)

The cut surface is represented by the two cross hatched tubes and a joining web. Note that the other of the two webs is not cross-hatched in accordance with drawing convention.

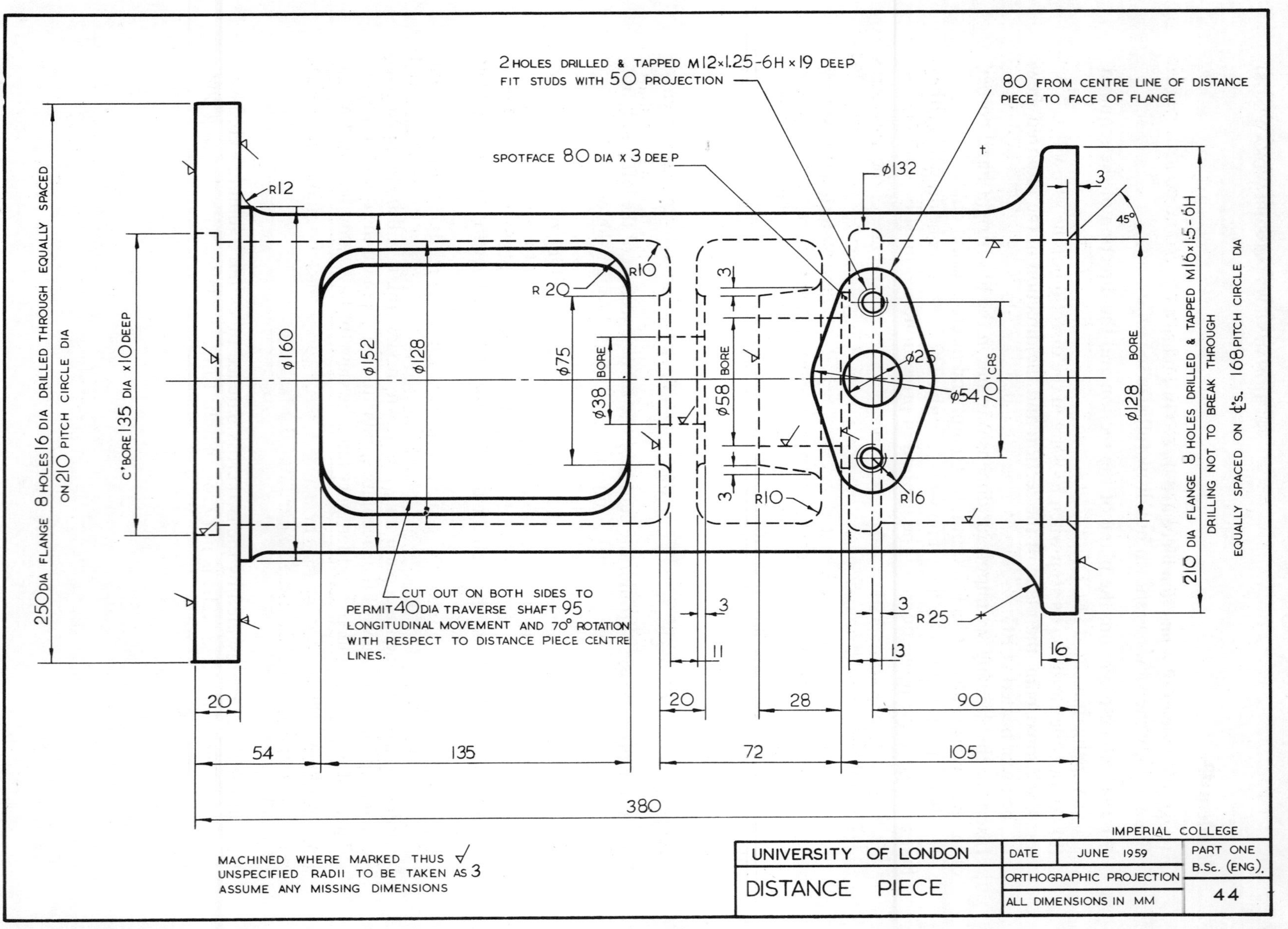

2 HOLES DRILLED & TAPPED M12×1.25-6H×19 DEEP
FIT STUDS WITH 50 PROJECTION
80 FROM CENTRE LINE OF DISTANCE PIECE TO FACE OF FLANGE
SPOTFACE 80 DIA x 3 DEEP
ϕ132
250 DIA FLANGE 8 HOLES 16 DIA DRILLED THROUGH EQUALLY SPACED
ON 210 PITCH CIRCLE DIA
C'BORE 135 DIA x10 DEEP
R12
R10
R 20
ϕ160
ϕ152
ϕ128
ϕ75
ϕ38 BORE
ϕ58 BORE
ϕ25
ϕ54
70 CRS
R16
ϕ128 BORE
45°
210 DIA FLANGE 8 HOLES DRILLED & TAPPED M16×1.5-6H
DRILLING NOT TO BREAK THROUGH
EQUALLY SPACED ON ℄s. 168 PITCH CIRCLE DIA
CUT OUT ON BOTH SIDES TO PERMIT 40 DIA TRAVERSE SHAFT 95 LONGITUDINAL MOVEMENT AND 70° ROTATION WITH RESPECT TO DISTANCE PIECE CENTRE LINES.
R 25
3
11
13
16
20
20
28
90
54
135
72
105
380
MACHINED WHERE MARKED THUS
UNSPECIFIED RADII TO BE TAKEN AS 3
ASSUME ANY MISSING DIMENSIONS
IMPERIAL COLLEGE
UNIVERSITY OF LONDON
DATE JUNE 1959
PART ONE B.Sc. (ENG).
ORTHOGRAPHIC PROJECTION
ALL DIMENSIONS IN MM
DISTANCE PIECE
44

Problem 44

A fully dimensioned drawing of a **distance piece** on a steam turbine control gear is shown in Figure 44.

Draw, half size:
(a) The outside view of the distance piece, as shown in Figure 44.
(b) Such other outside, or sectioned, views – projected from (a) or from one another – as may be necessary to give the complete shape and dimensions of the piece without the use of dotted lines for hidden parts. Use third angle projection; insert only those dimensions needed for fixing or location relative to other components.
Time allowed – $2\frac{1}{2}$ hours.

Solution (page 174)

A particularly interesting and useful exercise, it will by the very nature of the question's terminology result in a number of possible solutions. The solution shown on page 174 uses four complete views and an auxiliary part view for complete shape description. In fact, by using part views instead of complete views, a solution could be obtained more quickly than the one given. A more complete solution has been provided as an aid to visualisation.

Part (a)

A copy of one of the given views but without any hidden part lines. Note that the length of the cut out, i.e., 135 mm, is made up of the shaft's diameter – 40 mm – and the shaft's traverse – 95 mm. The width of the cut out cannot be determined until a view has been drawn which will show the effect of the 70° rotation of the shaft.

Part (b)

Section AA. This section is necessary in order to show complete shape description of the component's internal features. The cut away cannot be completed for similar reasons to those given in part (a).
Section BB. Although the near flange has been included in the model solution the sole reason for this view is to determine full shape description of the cut away. The partial auxiliary view serves to provide the true shape of the cut away corners and enables these corners to be projected into the previously mentioned views.
Half Section CC. The half section will show both studs assembled in the side flange. The examiner will be looking for correct stud proportions, and the student is advised to refer to the relevant British Standard on ISO fasteners.

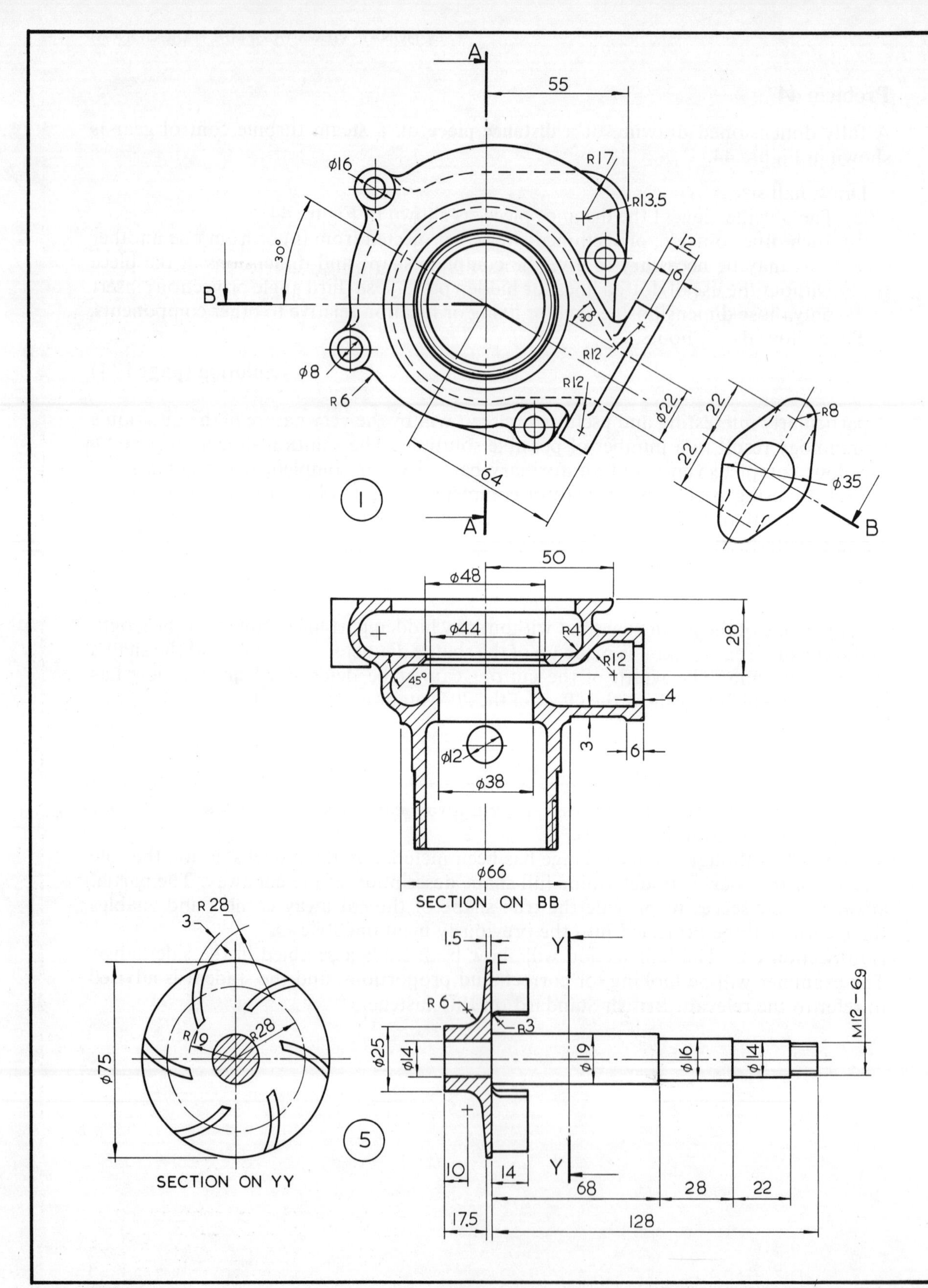

A
55
ϕ16
R17
R13.5
30°
12
6
B
30°
R12
ϕ8
R6
R12
ϕ22
22
22
R8
ϕ35
64
1
A
B
50
ϕ48
ϕ44
R4
28
R12
45°
4
3
6
ϕ12
ϕ38
ϕ66
SECTION ON BB
R 28
3
1.5
Y
F
R 6
R3
R19
R28
ϕ75
ϕ25
ϕ14
ϕ19
ϕ16
ϕ14
M12 – 6g
5
SECTION ON YY
10
14
Y
68
28
22
17.5
128

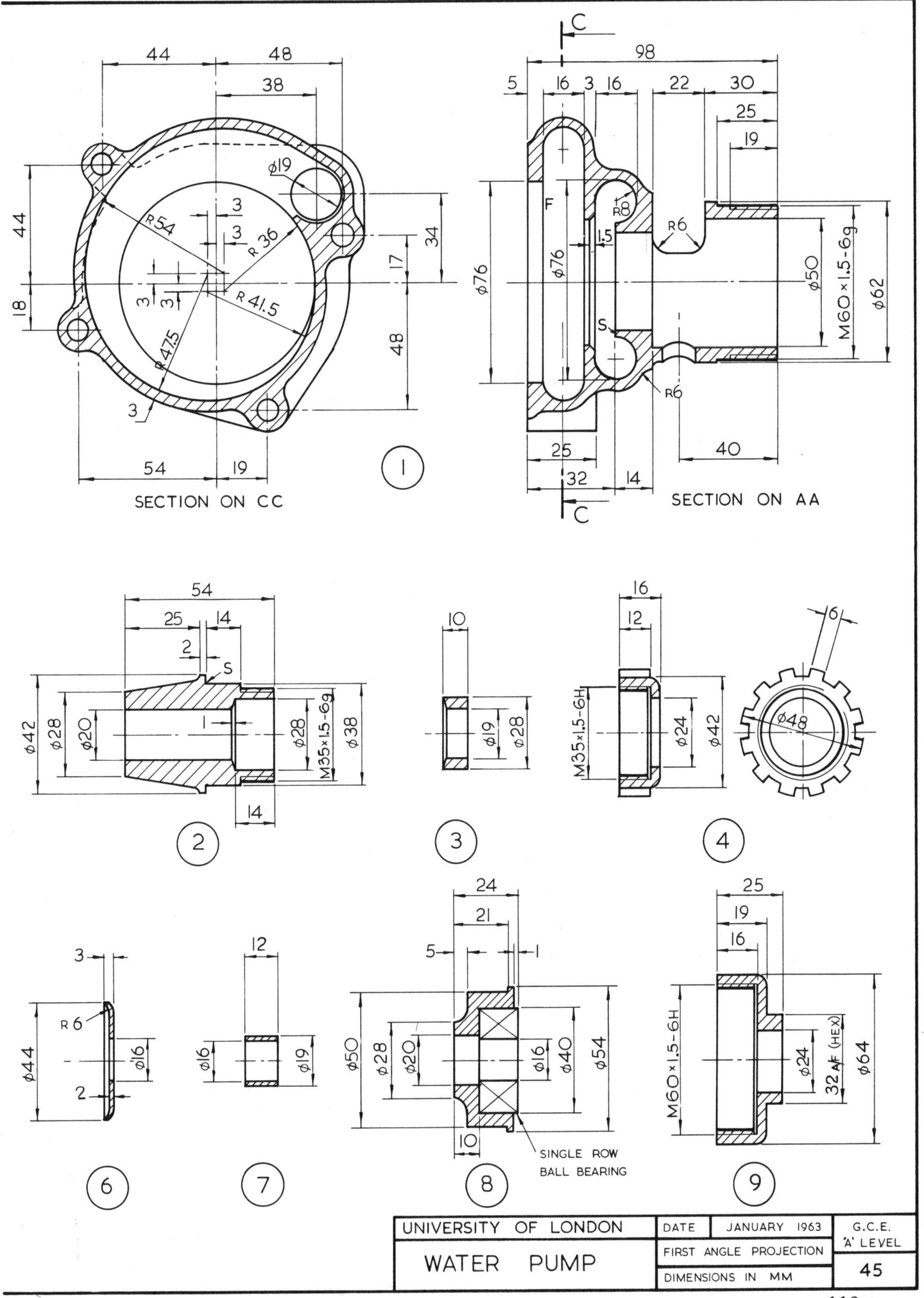

SECTION ON CC
SECTION ON AA
SINGLE ROW BALL BEARING
UNIVERSITY OF LONDON
DATE JANUARY 1963
G.C.E. 'A' LEVEL
WATER PUMP
FIRST ANGLE PROJECTION
DIMENSIONS IN MM
45

Problem 45

The detail components of a small **water circulating pump** are shown in Figure 45.

It is assembled with the combined bearing and gland (2) pressed into the body (1) so that the face S of the shoulder is in contact with the face S of the body. A circular slotted nut (4) engages the threaded diameter of (2) retaining the ring (3) in the gland.

The spindle (5) to which the impeller is attached passes through (2) and carries a water slinger (6) and a distance piece (7) on the reduced diameter of 15 mm. This portion of the spindle also has a single row ball bearing pressed on to it, thus securing (6) and (7) axially in position. The ball bearing is located in a housing (8) which is retained in the body (1) by means of the cap (9).

When the pump is assembled the face F of the impeller should be in line with the face F of the body, and the ring (3) entered 3 mm into the gland.

Draw to a scale of full size the following views of the assembled pump:

(a) A sectional end elevation on CC viewed in the direction of the arrows.
(b) A complete sectional elevation on AA as seen in the direction indicated.
(c) A complete outside plan view projected below (b).

View (a) must be drawn for purposes of projection, but need not be lined in or cross hatched.

Small unspecified radii should be taken as 3 mm.

Time allowed – 3 hours.

Solution (page 175)

View (a)

In order that the assembly can be built up gradually by drawing each component in its assembled position it is recommended that view (b) be concentrated on in the early stages. However, the main outline of the pump body – item 1 – needs to be drawn in view (a) in order that the diameter of the impeller housing can be projected into Section AA.

Careful compass work is necessary to produce the shape of the pump body in section CC. However, view (a) consists entirely of given views of items 1 and 5 and a simple section of item 2, and is more time-consuming than difficult.

View (b)

All parts on the question paper have been given in section through their centre line, and view (b) can be built up by reproducing the given views of the components in their respective positions. Matching faces of the main components have been given and this should assist considerably.

View (c)

The flange shown as an auxiliary view to item 1 on the question paper will be visible in view (c). Consequently the flange will need to be drawn lightly in its position in view (a) together with the auxiliary view in order that it may be projected into view (c). In fact the bulk of the work required in this view is involved in completing the shape description of item 1. The student would be well advised to complete the remainder of view (c) before spending a good deal of time interpreting the curves of intersection of item 1.

Problem 46

The details of the component parts of a **hand drill** are given in Figure 46.

Draw full size:

(a) An external elevation of the assembled parts.

(b) From this view project orthographically a 'sectional' elevation, taking the section along the longitudinal centre line. Consider the views carefully to ensure that those chosen will illustrate the object to the best advantage.

Items 8, 9 and 10, together with the chuck, items 14 to 18 inclusive, should be shown in external elevation only.

The chuck should be shown screwed along item 2 a distance of 19 mm.

Omit all dimensions and notation.

Use third angle projection.

Time allowed – 2 hours.

Solution (page 176)

Although no assembly instructions as such have been provided the examiners have helped by stating that items 14 to 18 inclusive constitute the chuck. Since the chuck is not required to be shown in section, items 15, 17 and 18 are not visible in either view. Each of the knobs and the handle have different spindle diameters which also helps in component identification. The student is advised to make a series of rough sketches as the position of each component becomes apparent. These can then be brought together to form a rough layout of the complete assembly.

Choice of views

The sectional view shown in the model solution enables all components to be shown in their working positions and the arrangement of the complex parts, i.e., items 4, 5 and 6, can be copied from the question paper. The viewing direction chosen for the outside elevation enables additional information to be given on the assembly of items 11, 12 and 13.

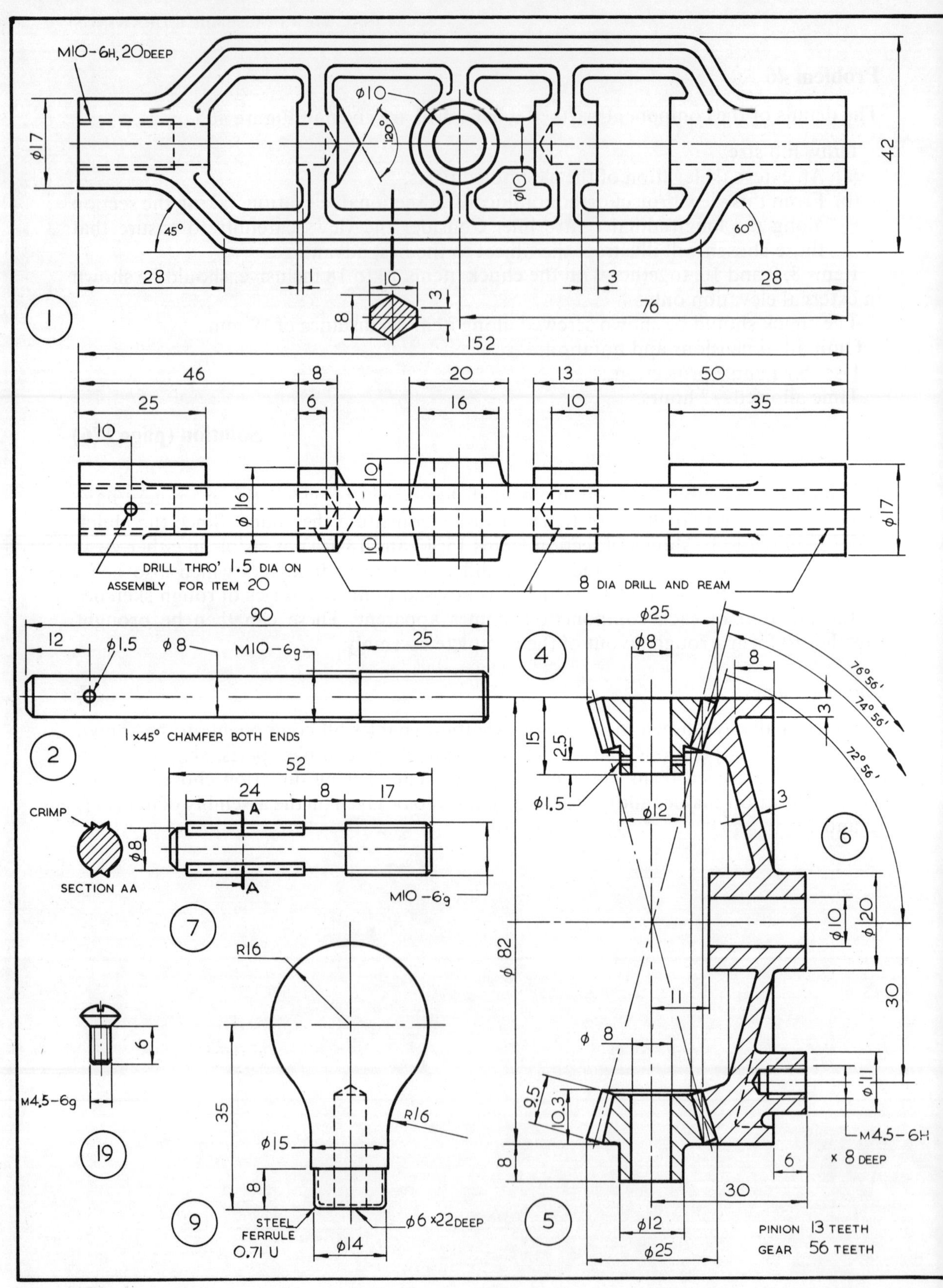

M10-6H, 20DEEP
φ10
120°
φ17
φ10
42
45°
60°
28
10
3
28
3
8
76
1
152
46
8
20
13
50
25
6
16
10
35
10
10
φ16
10
φ17
DRILL THRO' 1.5 DIA ON
ASSEMBLY FOR ITEM 20
8 DIA DRILL AND REAM
90
φ25
12
φ1.5
φ8
M10-6g
25
φ8
4
8
76°56'
74°56'
72°56'
3
1×45° CHAMFER BOTH ENDS
2
15
2.5
52
φ1.5
φ12
3
24
8
17
CRIMP
A
φ8
A
SECTION AA
6
M10-6g
φ10
φ20
7
φ82
R16
11
30
6
φ 8
M4.5-6g
9.5
10.3
35
R16
φ11
M4.5-6H
× 8 DEEP
19
φ15
8
6
8
30
STEEL
FERRULE
0.71 U
φ6 ×22DEEP
9
5
φ12
PINION 13 TEETH
GEAR 56 TEETH
φ14
φ25

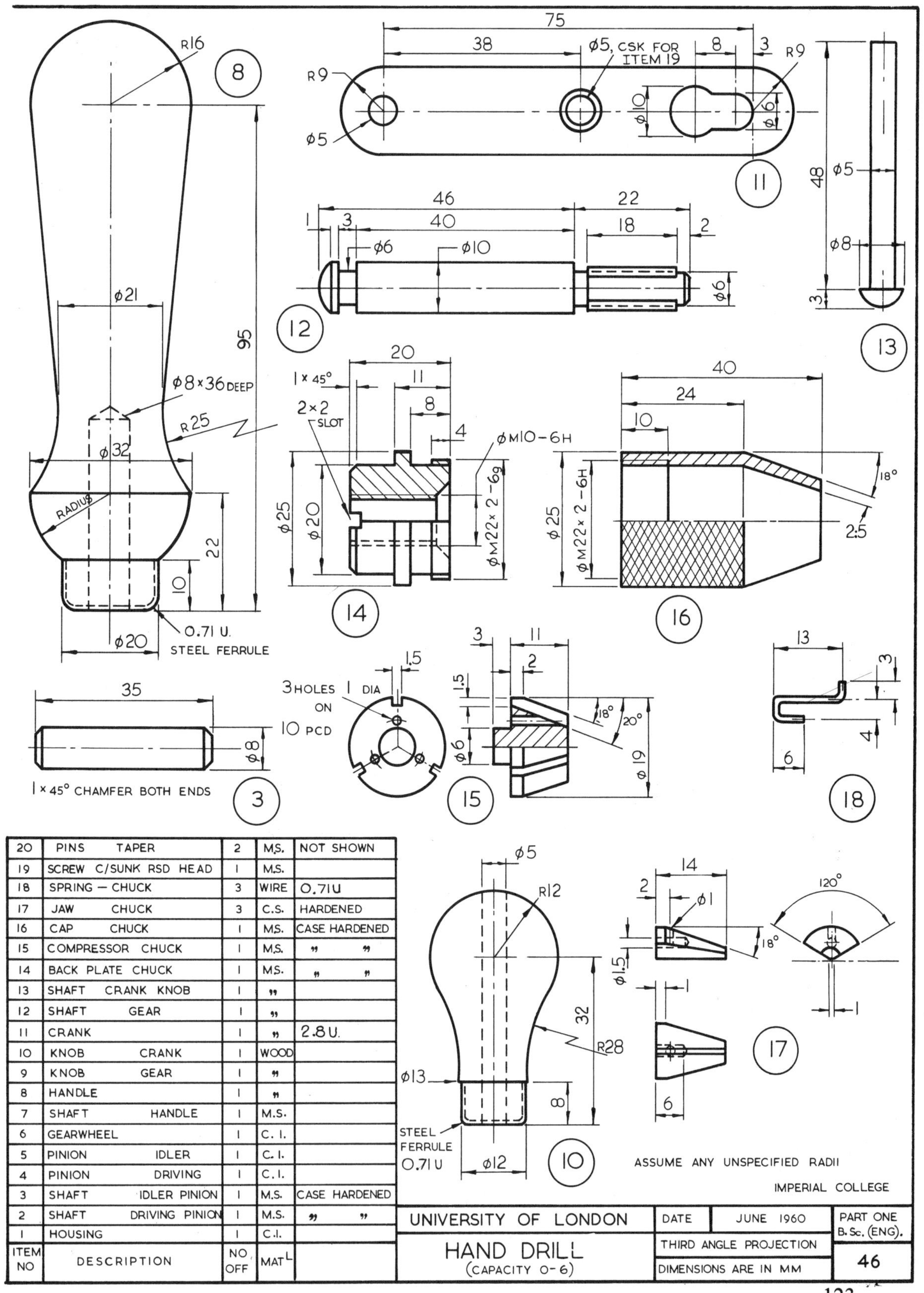

ITEM NO	DESCRIPTION	NO OFF	MAT^L	
20	PINS TAPER	2	M.S.	NOT SHOWN
19	SCREW C/SUNK RSD HEAD	1	M.S.	
18	SPRING – CHUCK	3	WIRE	0.71U
17	JAW CHUCK	3	C.S.	HARDENED
16	CAP CHUCK	1	M.S.	CASE HARDENED
15	COMPRESSOR CHUCK	1	M.S.	" "
14	BACK PLATE CHUCK	1	M.S.	" "
13	SHAFT CRANK KNOB	1	"	
12	SHAFT GEAR	1	"	
11	CRANK	1	"	2.8 U.
10	KNOB CRANK	1	WOOD	
9	KNOB GEAR	1	"	
8	HANDLE	1	"	
7	SHAFT HANDLE	1	M.S.	
6	GEARWHEEL	1	C. I.	
5	PINION IDLER	1	C. I.	
4	PINION DRIVING	1	C. I.	
3	SHAFT IDLER PINION	1	M.S.	CASE HARDENED
2	SHAFT DRIVING PINION	1	M.S.	" "
1	HOUSING	1	C.I.	

7 DRAWINGS FOR MANUFACTURE

Quite naturally the presentation of engineering components on many engineering drawing examination papers is concerned primarily with providing the candidate with a complete shape description of the part. It is not necessary in the majority of cases (the exception being some of the City and Guilds 293 examinations) for all the information for the part's manufacture to be included on the question paper. Typical items of information normally omitted include surface finish, heat treatment, material specification, etc. The following drawings, which include a number reproduced by kind permission of K and L Steelfounders and Engineers Ltd, are meant to illustrate the correct way to prepare drawings for manufacture, as recommended in BS 308.

Lever: Drawing Number K 30309

The first obvious difference from conventional examination question papers is in the layout of the drawing sheet. It is not sufficient to quote only the title, projection and scale, and consequently the drawing sheet must be designed to enable much more detailed information to be provided. Normally a company would study its whole range of products and list the various types of information that would have to be included on the component drawing. From this list a drawing sheet can be designed which will be suitable for all the company's products.

The design of the sheet on which the Lever is drawn includes provision for the following types of information.

A Material List
Normally only assembly drawings would require this block.

Modifications
From time to time, particularly during its early life, it may be necessary to impose some modification on the component's design. This may be due to the result of prototype testing which shows up a weakness in the design, and some improvement becomes necessary. Each modification is noted in the appropriate place on the drawing and a new issue number is allocated. It is then customary to provide the relevant production department with an up to date copy of the drawing and to ensure that the obsolete copy is destroyed.

Jigs and Fixtures
If a component is made in regular batches, and in sufficient quantities, jigs and fixtures may be made to help shorten machining time. If this is the case, then it will assist the machine operator considerably if the drawing lists the reference numbers of this equipment. He is then able to collect them from the stores together with the components that he has to machine.

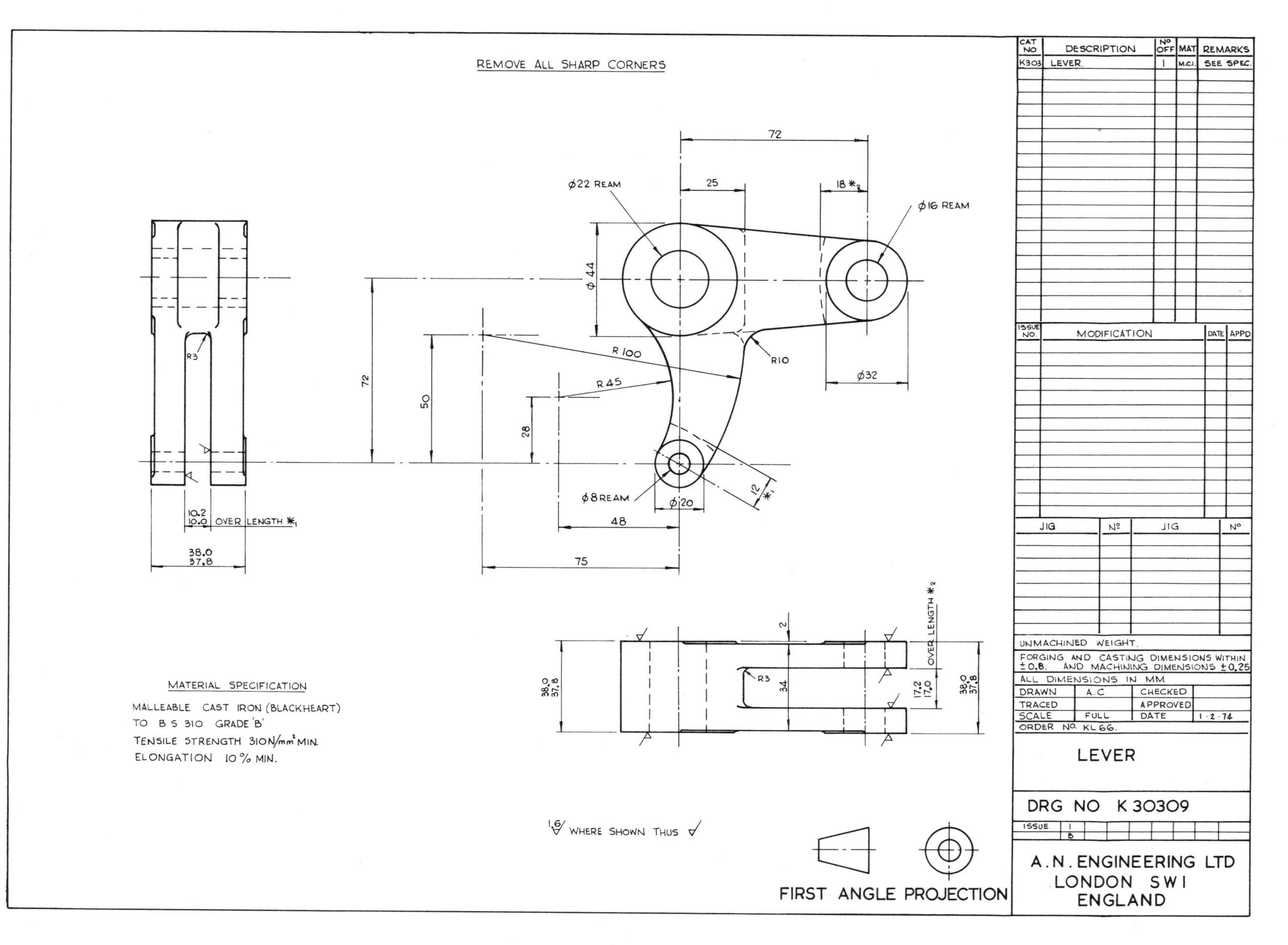
REMOVE ALL SHARP CORNERS
72
25
18 *2
ϕ22 REAM
ϕ16 REAM
ϕ 44
R3
R 100
R10
ϕ32
72
50
R 45
28
ϕ8 REAM
ϕ20
12 *1
48
75
10.2
10.0
OVER LENGTH *1
38.0
37.8
OVER LENGTH *2
2
R3
34
38.0
37.8
17.2
17.0
38.0
37.8
MATERIAL SPECIFICATION
MALLEABLE CAST IRON (BLACKHEART)
TO B S 310 GRADE 'B'
TENSILE STRENGTH 310N/mm² MIN.
ELONGATION 10 % MIN.
1.6 WHERE SHOWN THUS
FIRST ANGLE PROJECTION
CAT NO
DESCRIPTION
Nº OFF
MAT
REMARKS
K303
LEVER.
1
M.C.I.
SEE SPEC.
ISSUE NO.
MODIFICATION
DATE
APPD
JIG
Nº
JIG
Nº
UNMACHINED WEIGHT.
FORGING AND CASTING DIMENSIONS WITHIN ±0.8. AND MACHINING DIMENSIONS ±0.25
ALL DIMENSIONS IN MM
DRAWN
A.C
CHECKED
TRACED
APPROVED
SCALE
FULL
DATE
1.2.74
ORDER NO. KL 66.
LEVER
DRG NO K 30309
ISSUE
1
B
A. N. ENGINEERING LTD
LONDON SW1
ENGLAND

Weight
This is a necessary item of information where larger components are concerned. The planning department will need to ensure, for instance, that suitable handling equipment is available for the transportation of the part to the machine tool and for the component's manipulation on perhaps a jig borer or lathe.

General Tolerances
It is impossible to manufacture components to their exact theoretical design sizes by economic production methods, and consequently every dimension given on an engineering drawing must be provided with a tolerance; the amount of deviation from the nominal size.

Normally all dimensions will be subject to a general tolerance, the value of which will depend upon the class of product and the type of operation producing that dimension. The general tolerances quoted on Drawing Number K 30309 are $\pm 0{\cdot}8$ mm on forged or cast dimensions and $\pm 0{\cdot}25$ mm on machined dimensions.

For the purpose of interchangeability it may be necessary to enforce more specific tolerances. Examples of these are shown on the drawing of the Lever, i.e., 38·0 mm, 37·8 mm. For further information on this topic the student is referred to BS 4500 and BS 1916: Part 2.

Order Number
Many companies include all or part of the order reference number as part of the drawing number. If not, it is included as a further useful form of reference.

Material Specification
It is not sufficiently accurate to merely quote 'mild steel' or 'cast iron' as a material description. There are a considerable number of mild steels and cast irons available, all of which have a British Standard reference number. A list of the numbers for steels can be found in BS 970 together with each material's composition and physical characteristics. A study of this standard will enable the correct material to be selected and stated on the drawing.

Heat Treatment
Where this is necessary, full details must be given, i.e., temperatures and quenching media.

Surface Finish
All machined surfaces can have varying degrees of smoothness depending on the method and tool chosen to produce it. Where it is important that a surface has a particular surface finish, a symbol is used together with an arrangement of numbers, symbols and letters:
e.g.

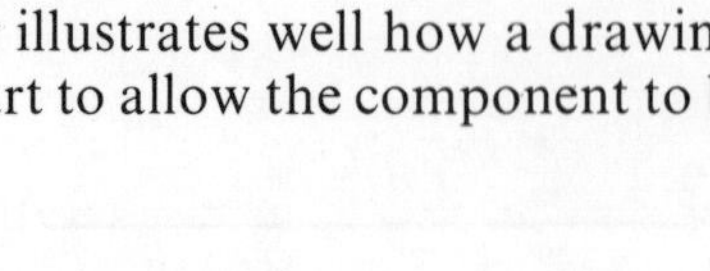

1.6 M or 0.8 ⊥

Further information relating to surface texture may be obtained from BS 1134.

General Comments
The drawing of the Lever illustrates well how a drawing should be laid out. The views are spread far enough apart to allow the component to be dimensioned clearly and with

an uncramped style. One of the major faults of the inexperienced draughtsman is to draw dimension lines too close together.

Clutch Centre Plate: Drawing Number K 6341

A similar drawing sheet has been used to that for the Lever and as a result it is possible to provide full details of the component.

General Comments

Details of the component in the forged condition have been included on the machining drawing as an aid to clarity.

Bevel Pinion: Drawing Number M 608

The drawing sheet is similar to that used in the two previous examples. It is interesting to read the types of modification that could occur in the life of a component. In this case a major change – from a machined blank to a stamping. As in the previous drawing, the form in which the component is initially produced has been included for clarity, i.e., a stamping. Although a considerable amount of information has been given on the drawing sheet, the neatness of the layout enables all features to be identified clearly.

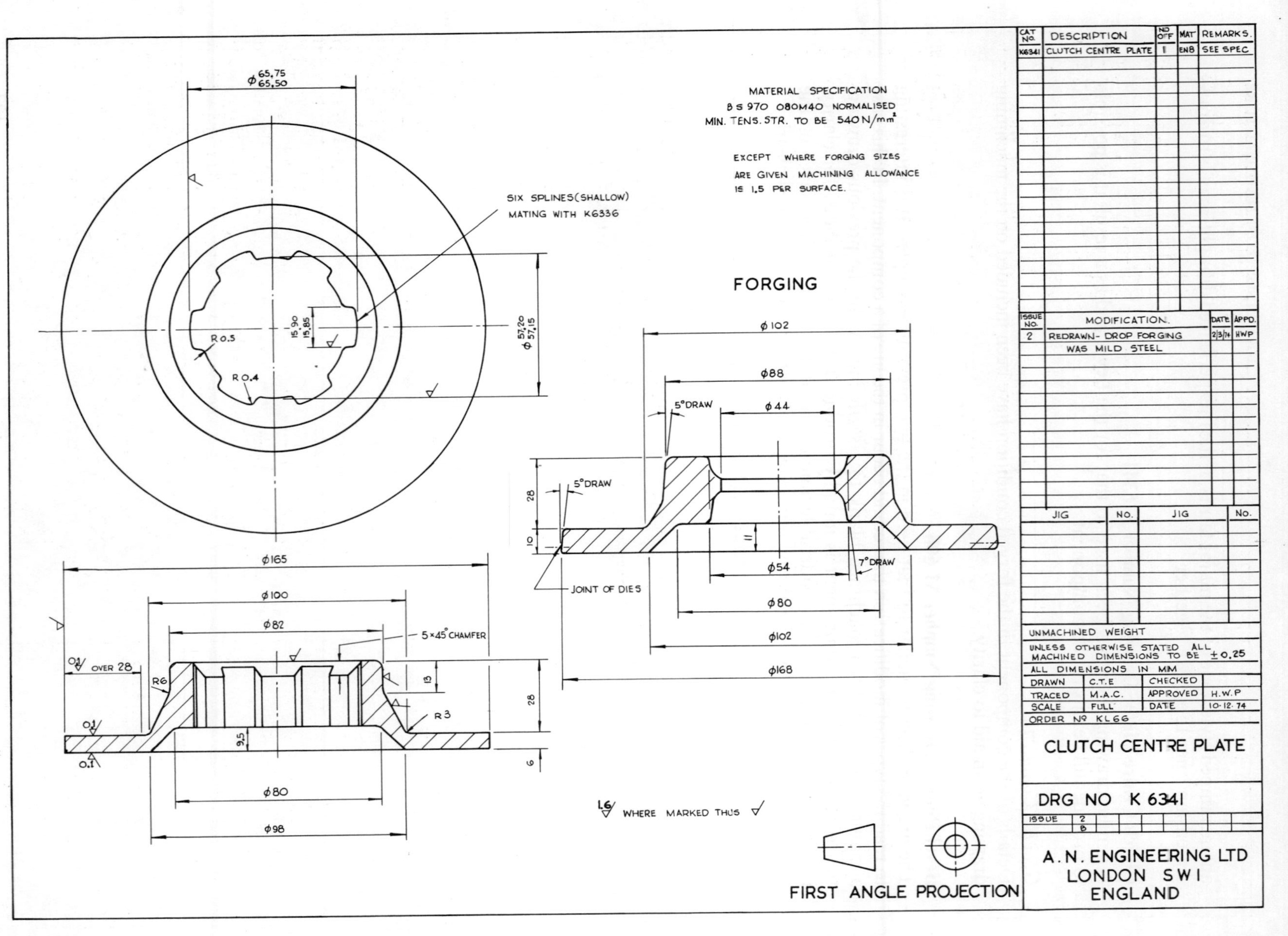
MATERIAL SPECIFICATION
BS 970 080M40 NORMALISED
MIN. TENS. STR. TO BE 540 N/mm²
EXCEPT WHERE FORGING SIZES
ARE GIVEN MACHINING ALLOWANCE
IS 1,5 PER SURFACE.
SIX SPLINES (SHALLOW)
MATING WITH K6336
FORGING
JOINT OF DIES
5°DRAW
7°DRAW
5×45° CHAMFER
1.6 WHERE MARKED THUS
FIRST ANGLE PROJECTION
CAT No. | DESCRIPTION | No OFF | MAT | REMARKS
K6341 | CLUTCH CENTRE PLATE | 1 | EN8 | SEE SPEC
ISSUE No. | MODIFICATION | DATE | APPD.
2 | REDRAWN- DROP FORGING WAS MILD STEEL | 2/3/74 | HWP
JIG | No. | JIG | No.
UNMACHINED WEIGHT
UNLESS OTHERWISE STATED ALL MACHINED DIMENSIONS TO BE ±0,25
ALL DIMENSIONS IN MM
DRAWN C.T.E CHECKED
TRACED M.A.C. APPROVED H.W.P
SCALE FULL DATE 10-12-74
ORDER Nº KL66
CLUTCH CENTRE PLATE
DRG NO K 6341
ISSUE 2 B
A.N. ENGINEERING LTD
LONDON SW1
ENGLAND

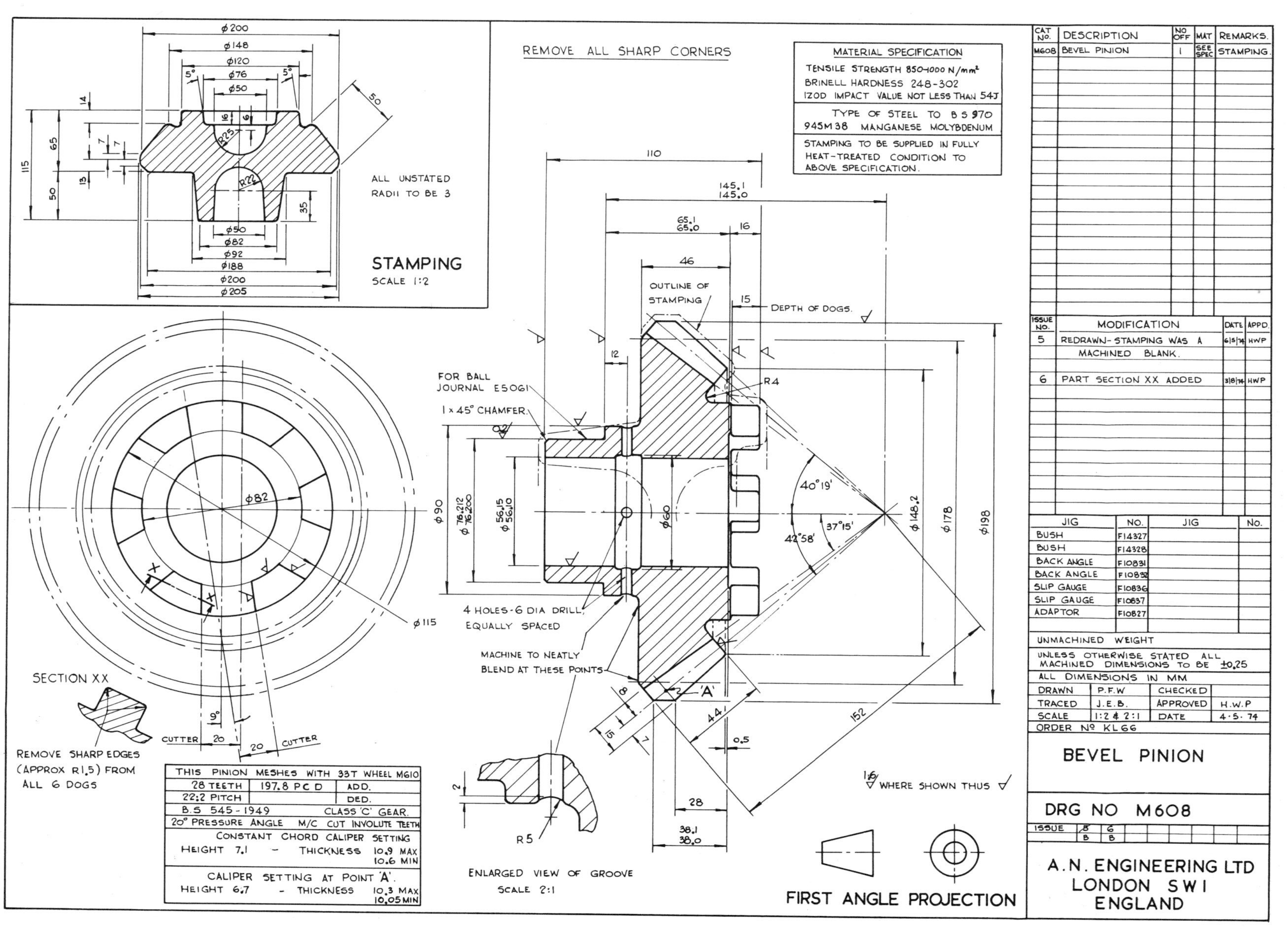
REMOVE ALL SHARP CORNERS
MATERIAL SPECIFICATION
TENSILE STRENGTH 850-1000 N/mm²
BRINELL HARDNESS 248-302
IZOD IMPACT VALUE NOT LESS THAN 54J
TYPE OF STEEL TO B S 970
945M38 MANGANESE MOLYBDENUM
STAMPING TO BE SUPPLIED IN FULLY HEAT-TREATED CONDITION TO ABOVE SPECIFICATION.
ALL UNSTATED RADII TO BE 3
STAMPING
SCALE 1:2
OUTLINE OF STAMPING
DEPTH OF DOGS.
FOR BALL JOURNAL E50G1
1 x 45° CHAMFER.
4 HOLES-6 DIA DRILL, EQUALLY SPACED
MACHINE TO NEATLY BLEND AT THESE POINTS
SECTION XX
REMOVE SHARP EDGES (APPROX R1.5) FROM ALL 6 DOGS
CUTTER
ENLARGED VIEW OF GROOVE
SCALE 2:1
WHERE SHOWN THUS
FIRST ANGLE PROJECTION
THIS PINION MESHES WITH 33T WHEEL M610
28 TEETH 197.8 P C D ADD.
22.2 PITCH DED.
B.S 545-1949 CLASS 'C' GEAR.
20° PRESSURE ANGLE M/C CUT INVOLUTE TEETH
CONSTANT CHORD CALIPER SETTING
HEIGHT 7.1 - THICKNESS 10.9 MAX 10.6 MIN
CALIPER SETTING AT POINT 'A'.
HEIGHT 6.7 - THICKNESS 10.3 MAX 10.05 MIN
CAT NO. DESCRIPTION NO OFF MAT REMARKS.
M608 BEVEL PINION 1 SEE SPEC STAMPING.
ISSUE NO. MODIFICATION DATE APPD.
5 REDRAWN- STAMPING WAS A MACHINED BLANK. 6/5/74 HWP
6 PART SECTION XX ADDED 3/8/74 HWP
JIG NO. JIG No.
BUSH F14327
BUSH F14328
BACK ANGLE F10831
BACK ANGLE F10832
SLIP GAUGE F10836
SLIP GAUGE F10837
ADAPTOR F10827
UNMACHINED WEIGHT
UNLESS OTHERWISE STATED ALL MACHINED DIMENSIONS TO BE ±0.25
ALL DIMENSIONS IN MM
DRAWN P.F.W CHECKED
TRACED J.E.B. APPROVED H.W.P
SCALE 1:2 & 2:1 DATE 4-5-74
ORDER Nº KL66
BEVEL PINION
DRG NO M608
ISSUE 5 6
A.N. ENGINEERING LTD
LONDON SW1
ENGLAND

SOLUTIONS

List of Solutions to Problems

The solutions following are complete except that neither dimensions nor scale have been shown. Methods of indicating these are shown in Chapter 7 on 'Drawings for Manufacture'.

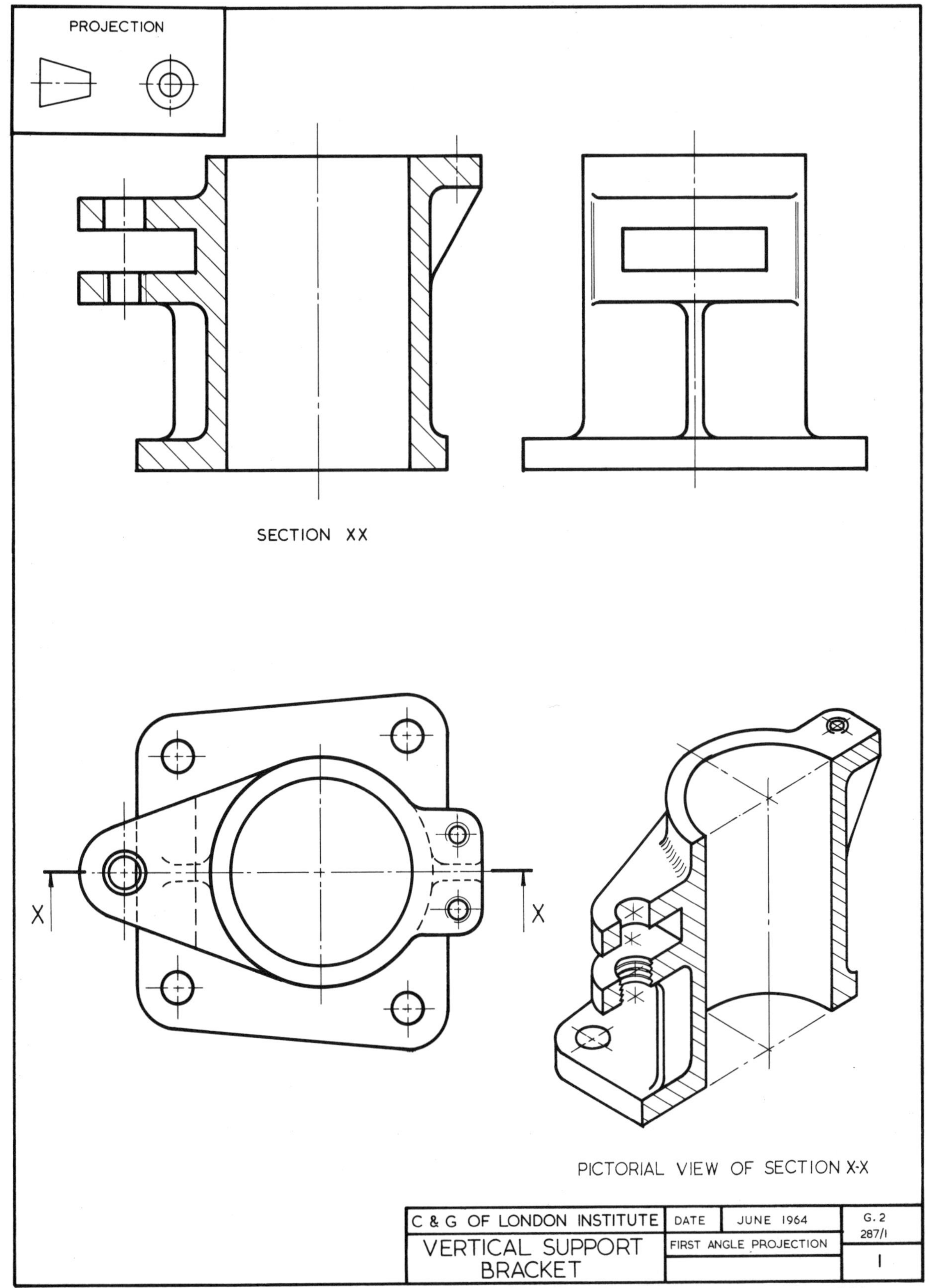
PROJECTION
SECTION XX
X
X
PICTORIAL VIEW OF SECTION X-X
C & G OF LONDON INSTITUTE
DATE
JUNE 1964
G. 2
287/1
VERTICAL SUPPORT
BRACKET
FIRST ANGLE PROJECTION
1

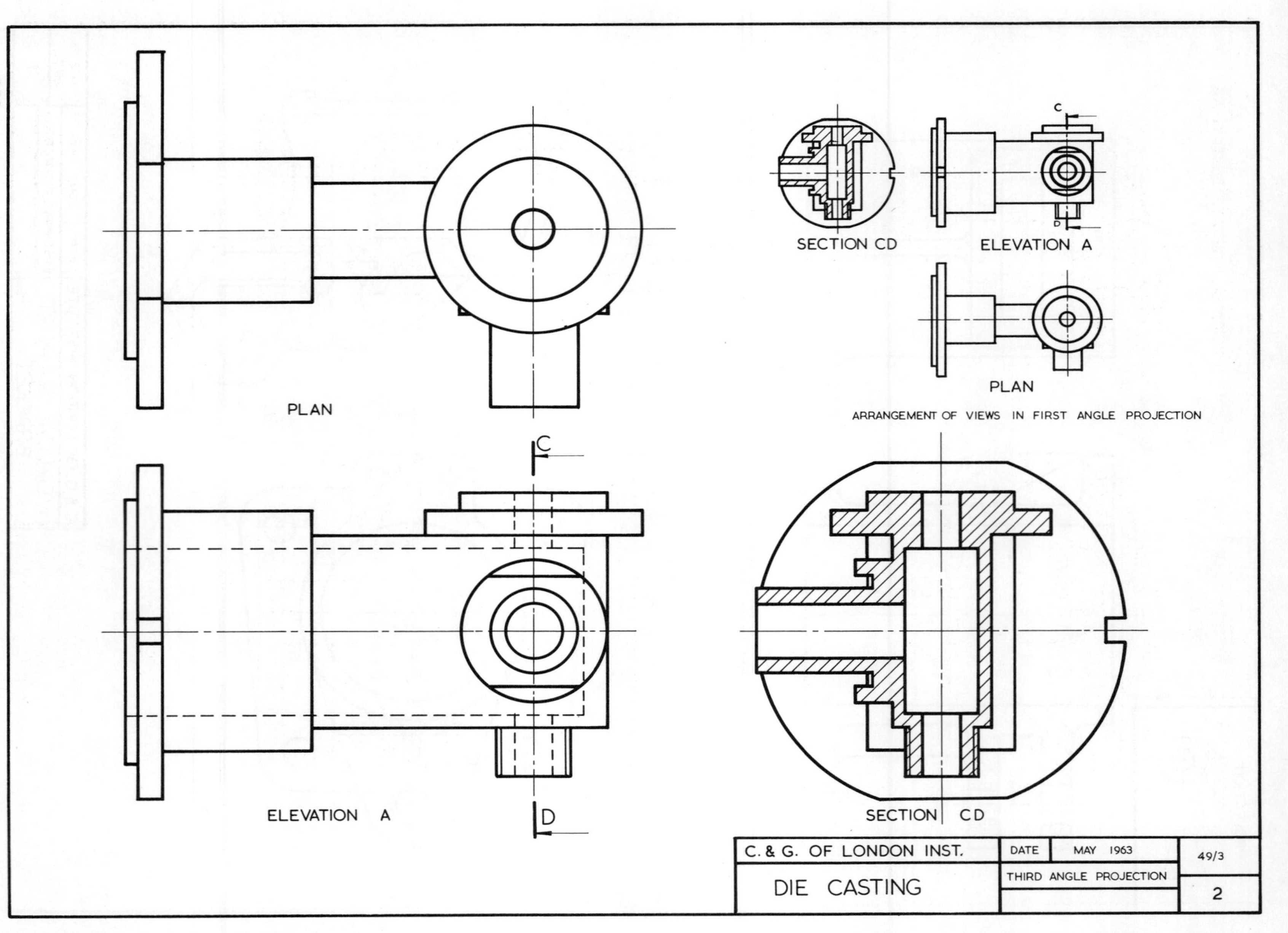
SECTION CD
ELEVATION A
C
D
PLAN
ARRANGEMENT OF VIEWS IN FIRST ANGLE PROJECTION
PLAN
C
ELEVATION A
D
SECTION CD
C. & G. OF LONDON INST.
DATE
MAY 1963
49/3
DIE CASTING
THIRD ANGLE PROJECTION
2

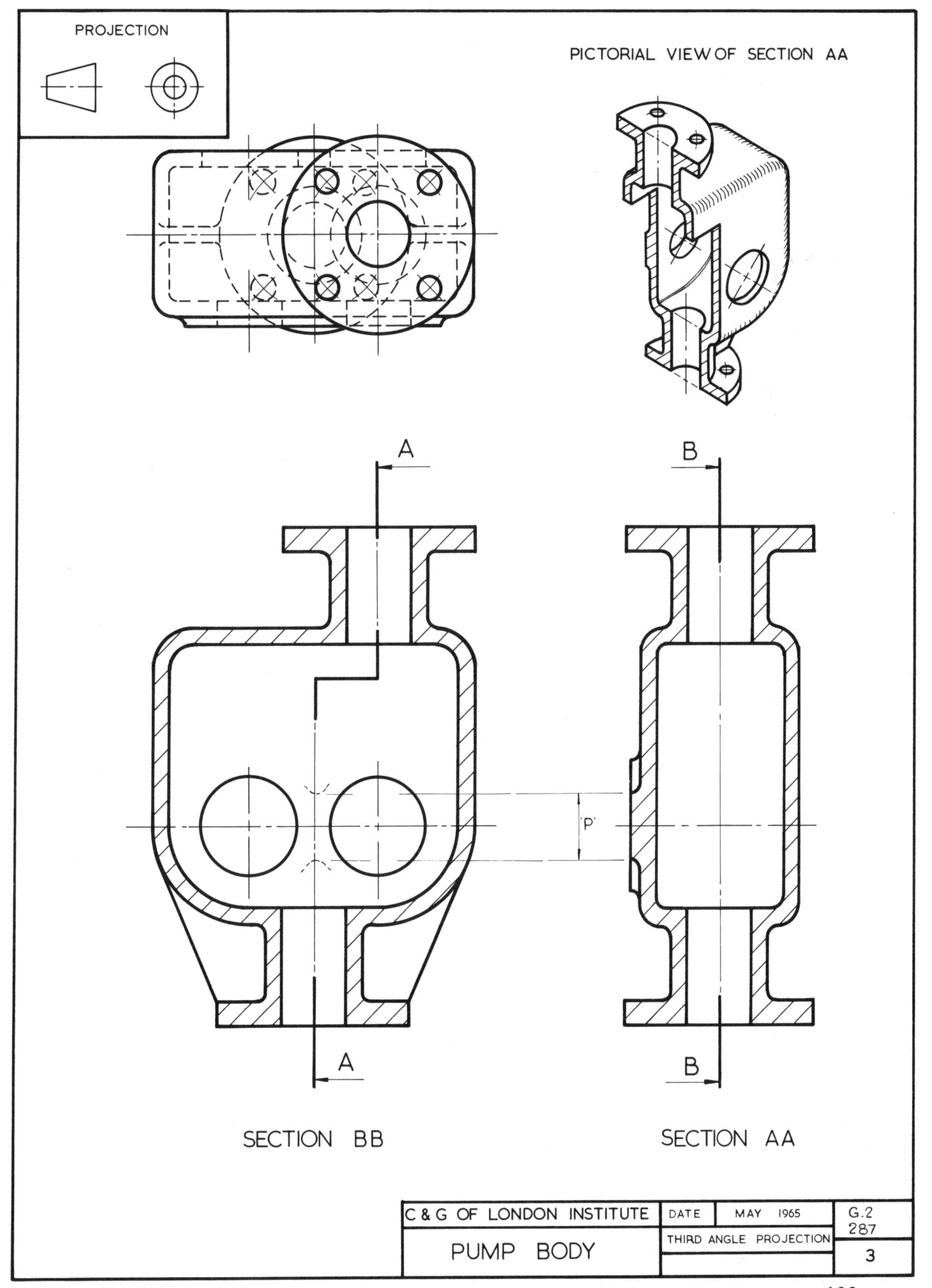
PROJECTION
PICTORIAL VIEW OF SECTION AA
A
B
P
A
B
SECTION BB
SECTION AA
C & G OF LONDON INSTITUTE
DATE
MAY 1965
G.2
287
PUMP BODY
THIRD ANGLE PROJECTION
3

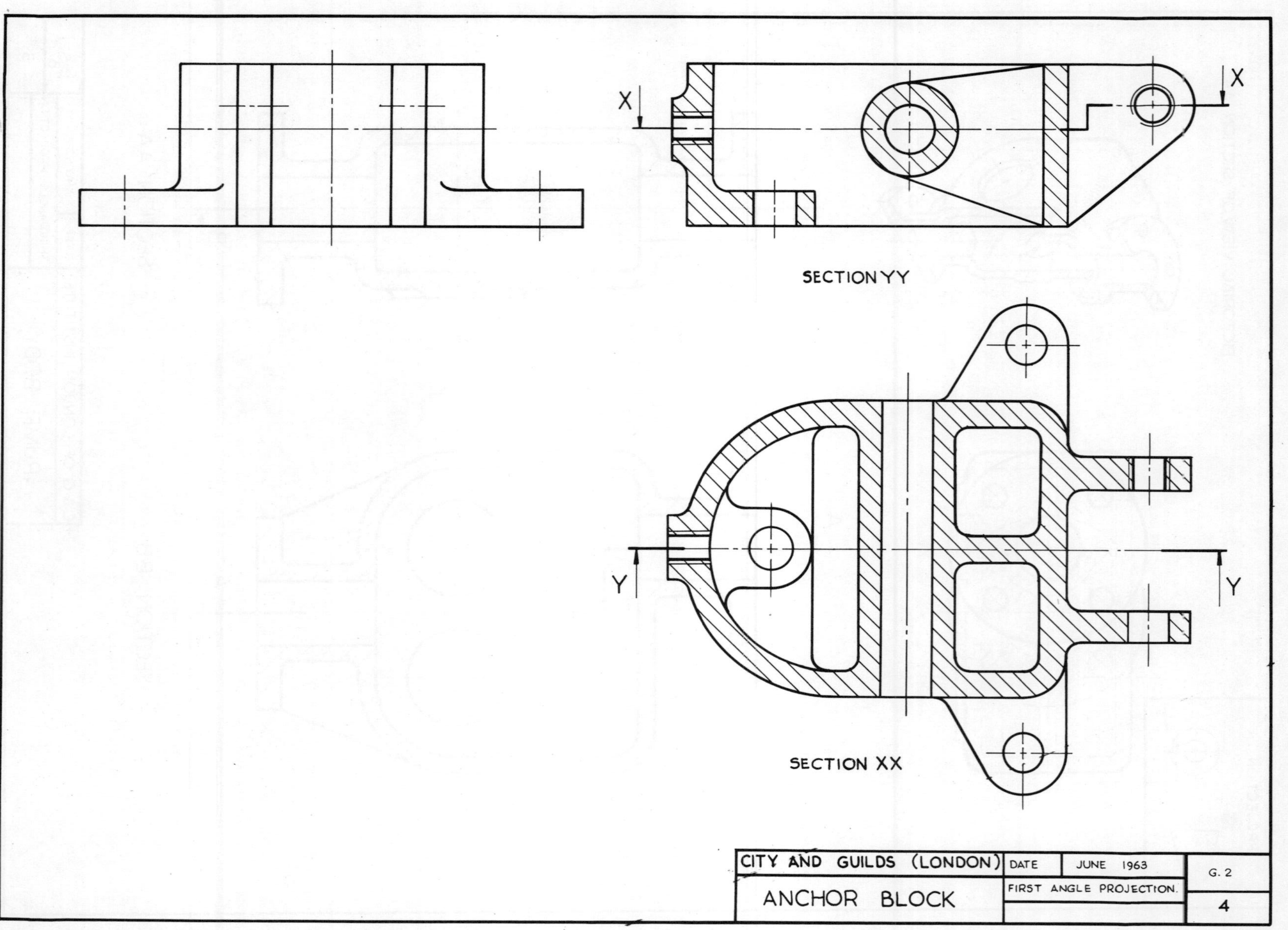
X
X
SECTION YY
Y
Y
SECTION XX
CITY AND GUILDS (LONDON)
DATE
JUNE 1963
G. 2
ANCHOR BLOCK
FIRST ANGLE PROJECTION.
4

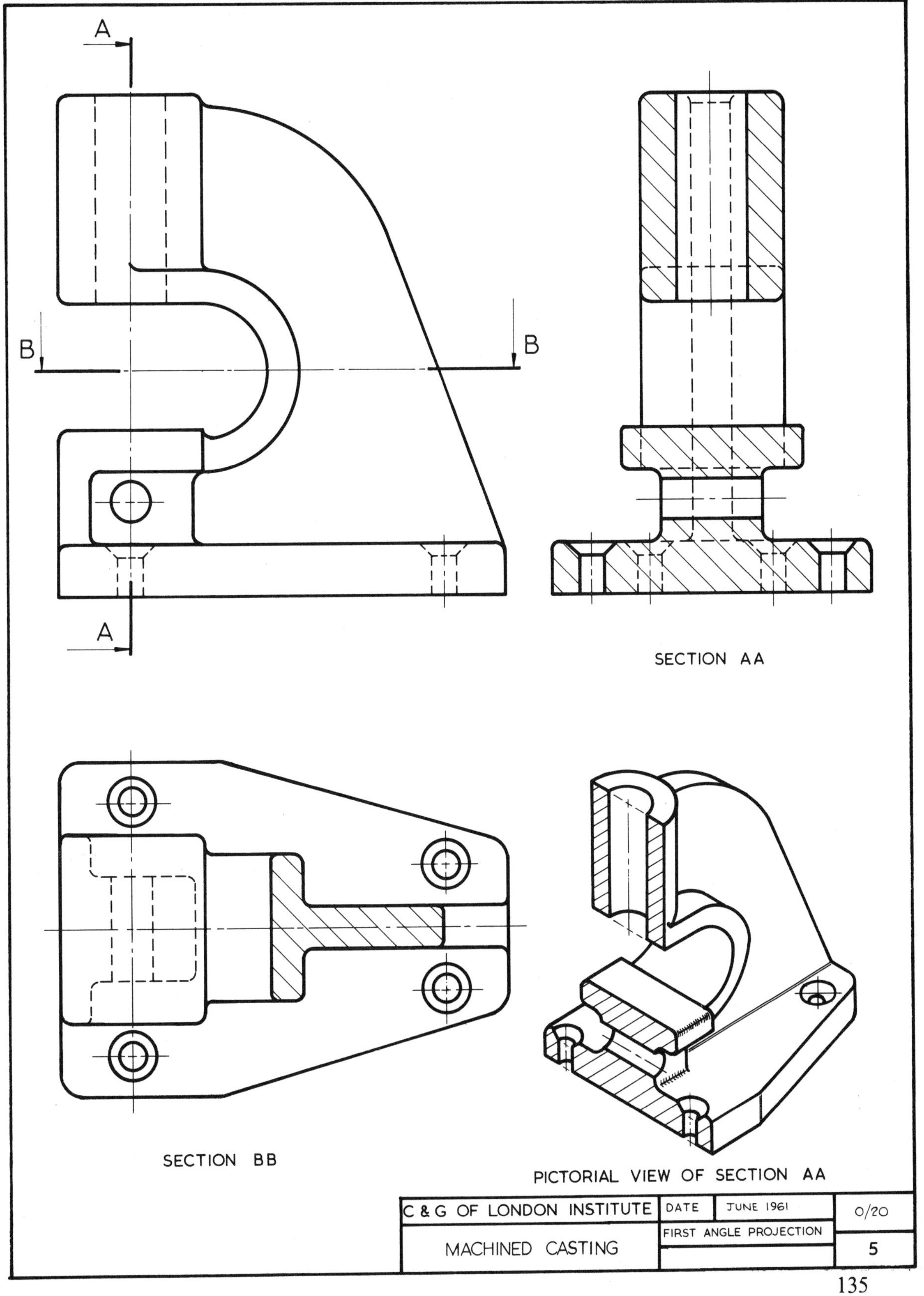
A
A
B
B
SECTION AA
SECTION BB
PICTORIAL VIEW OF SECTION AA
C & G OF LONDON INSTITUTE
DATE
JUNE 1961
0/20
MACHINED CASTING
FIRST ANGLE PROJECTION
5

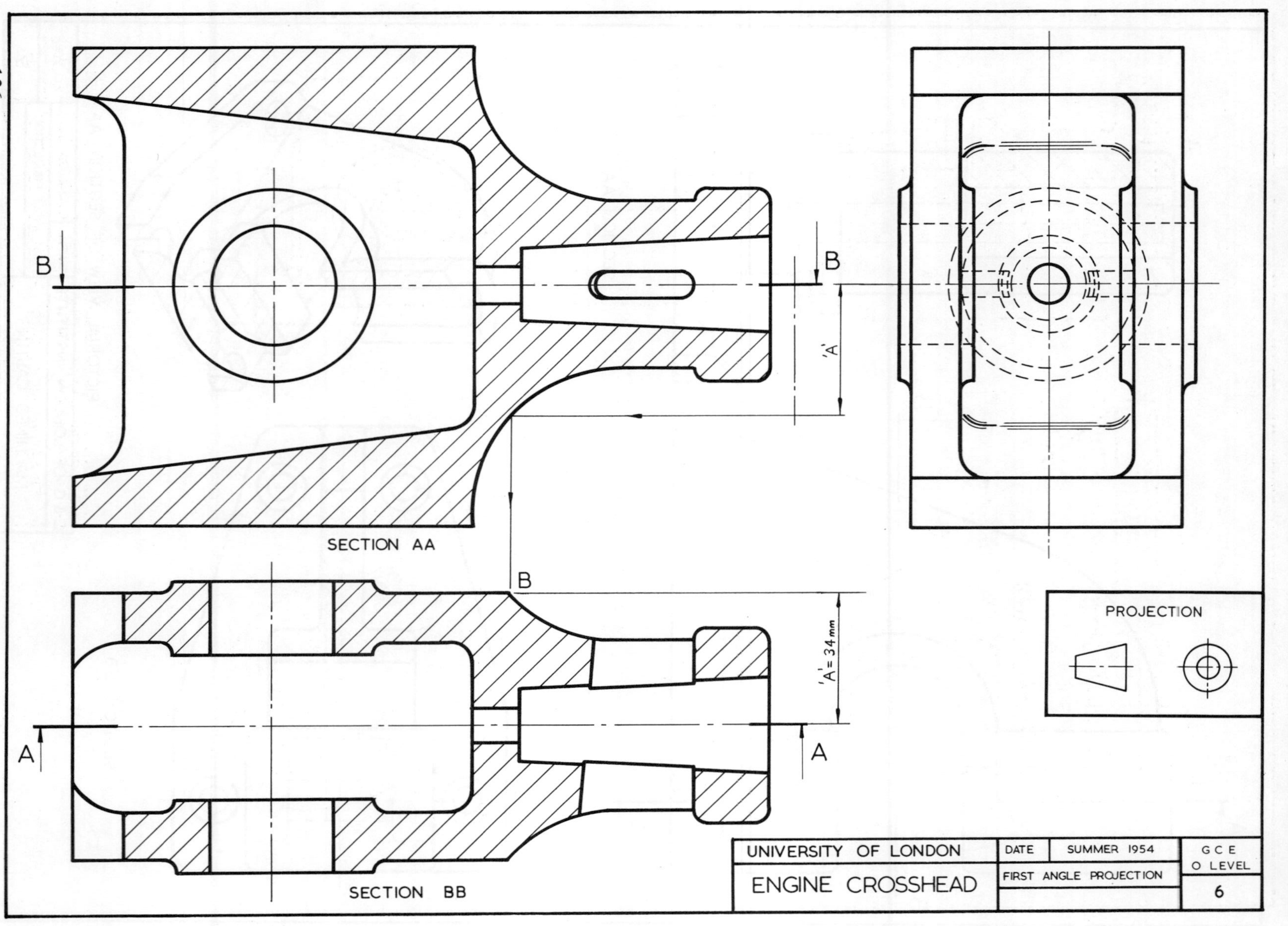

B
B
'A'
SECTION AA
B
'A' = 34 mm
A
A
PROJECTION
SECTION BB
UNIVERSITY OF LONDON
ENGINE CROSSHEAD
DATE
SUMMER 1954
FIRST ANGLE PROJECTION
G C E
O LEVEL
6

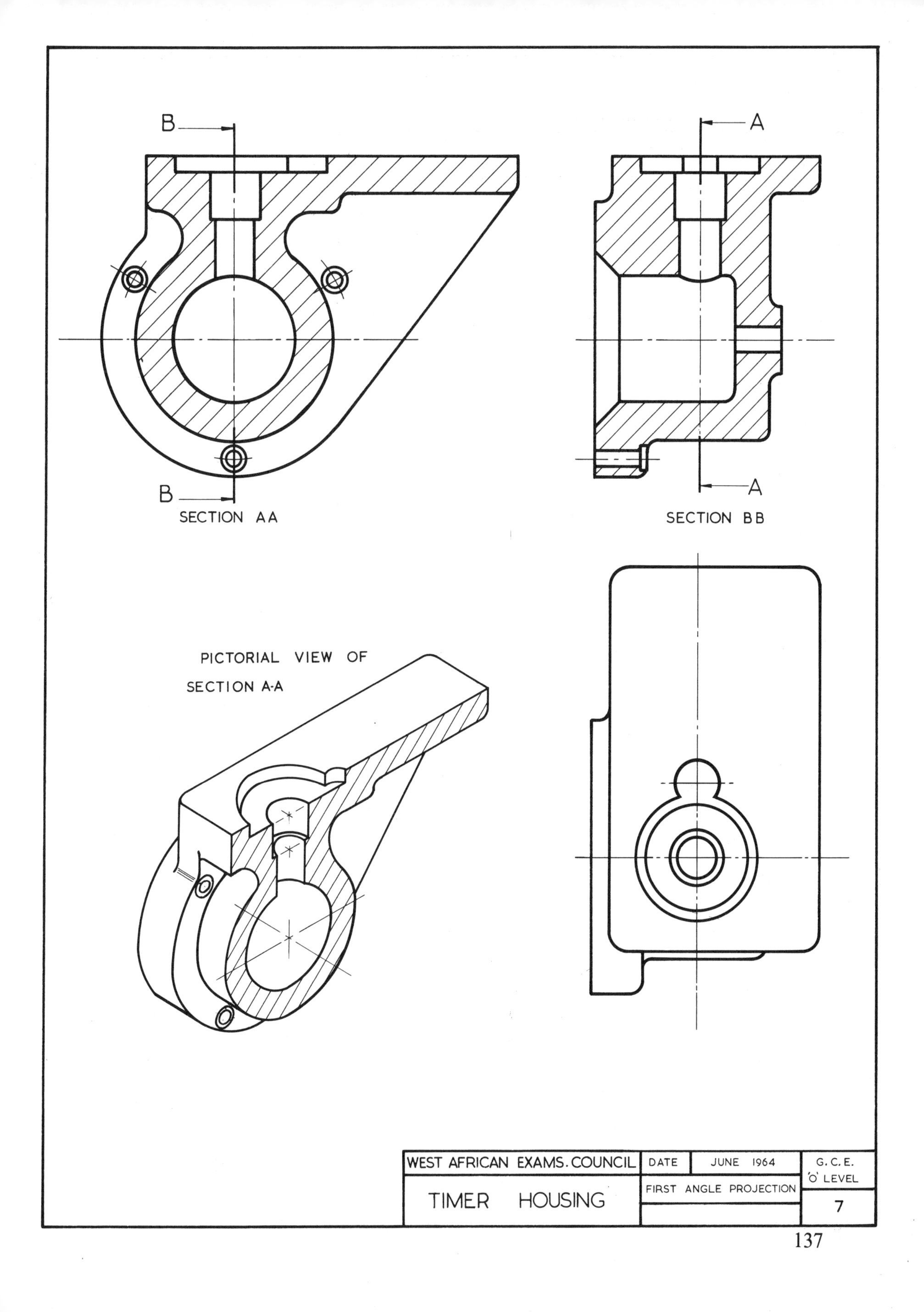
B
B
SECTION AA
A
A
SECTION BB
PICTORIAL VIEW OF
SECTION A-A
WEST AFRICAN EXAMS. COUNCIL
DATE
JUNE 1964
G. C. E.
'O' LEVEL
TIMER HOUSING
FIRST ANGLE PROJECTION
7

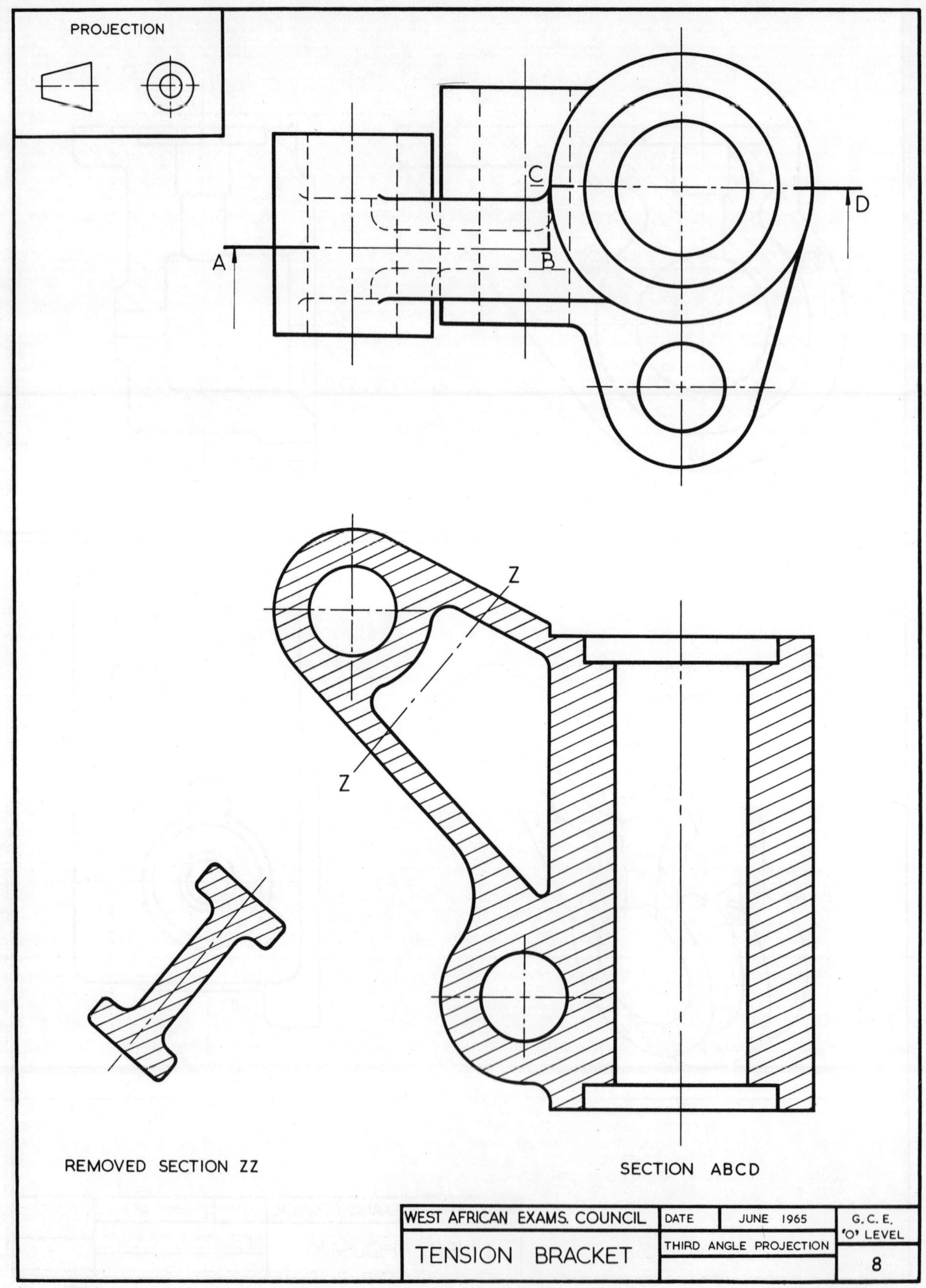
PROJECTION
C
D
A
B
Z
Z
REMOVED SECTION ZZ
SECTION ABCD
WEST AFRICAN EXAMS. COUNCIL
DATE
JUNE 1965
G. C. E.
'O' LEVEL
TENSION BRACKET
THIRD ANGLE PROJECTION
8

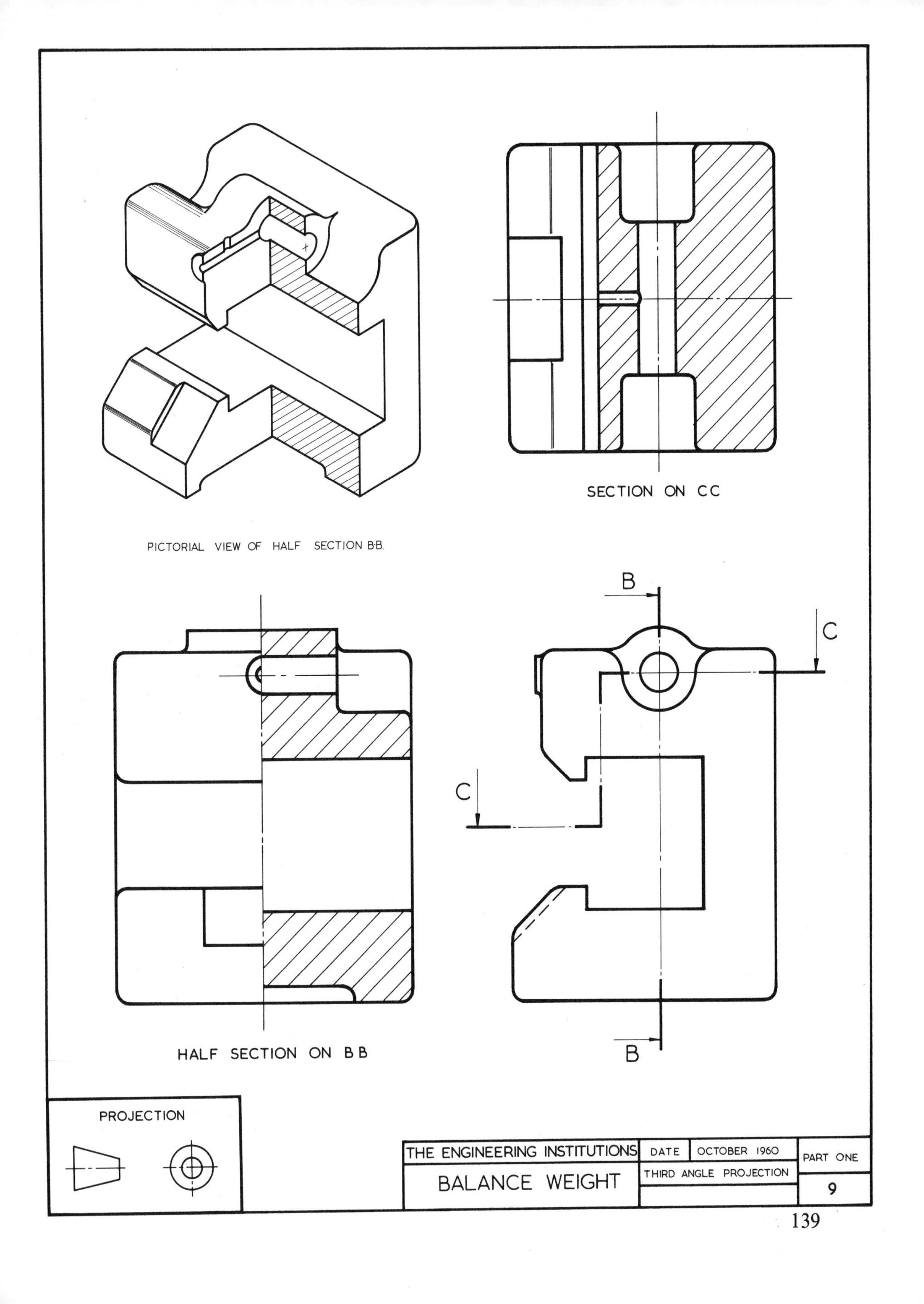
PICTORIAL VIEW OF HALF SECTION B-B.
SECTION ON CC
B
C
C
B
HALF SECTION ON BB
PROJECTION
THE ENGINEERING INSTITUTIONS
DATE
OCTOBER 1960
PART ONE
BALANCE WEIGHT
THIRD ANGLE PROJECTION
9

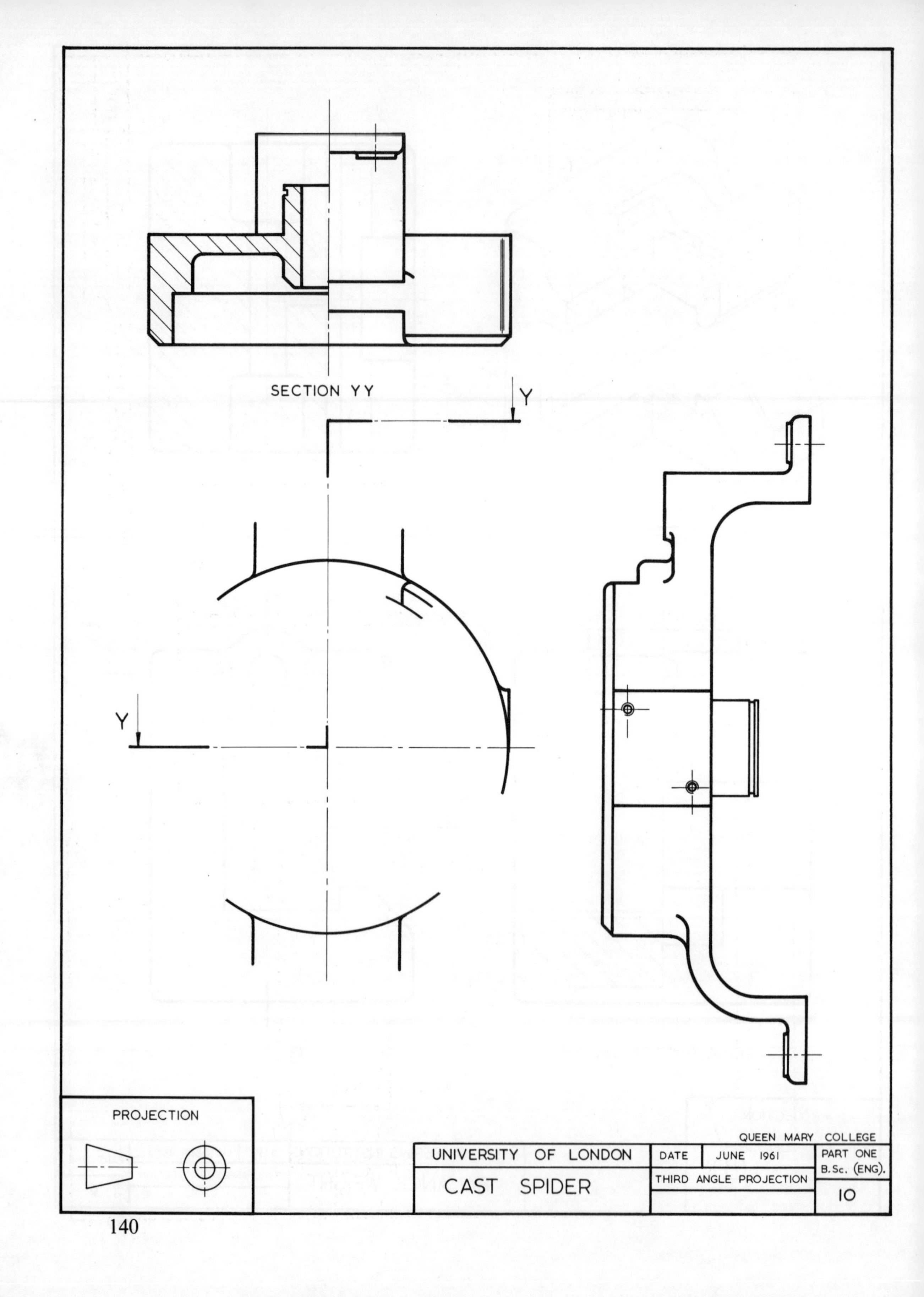
SECTION YY
Y
Y
PROJECTION
QUEEN MARY COLLEGE
UNIVERSITY OF LONDON
DATE
JUNE 1961
PART ONE
B.Sc. (ENG).
CAST SPIDER
THIRD ANGLE PROJECTION
10

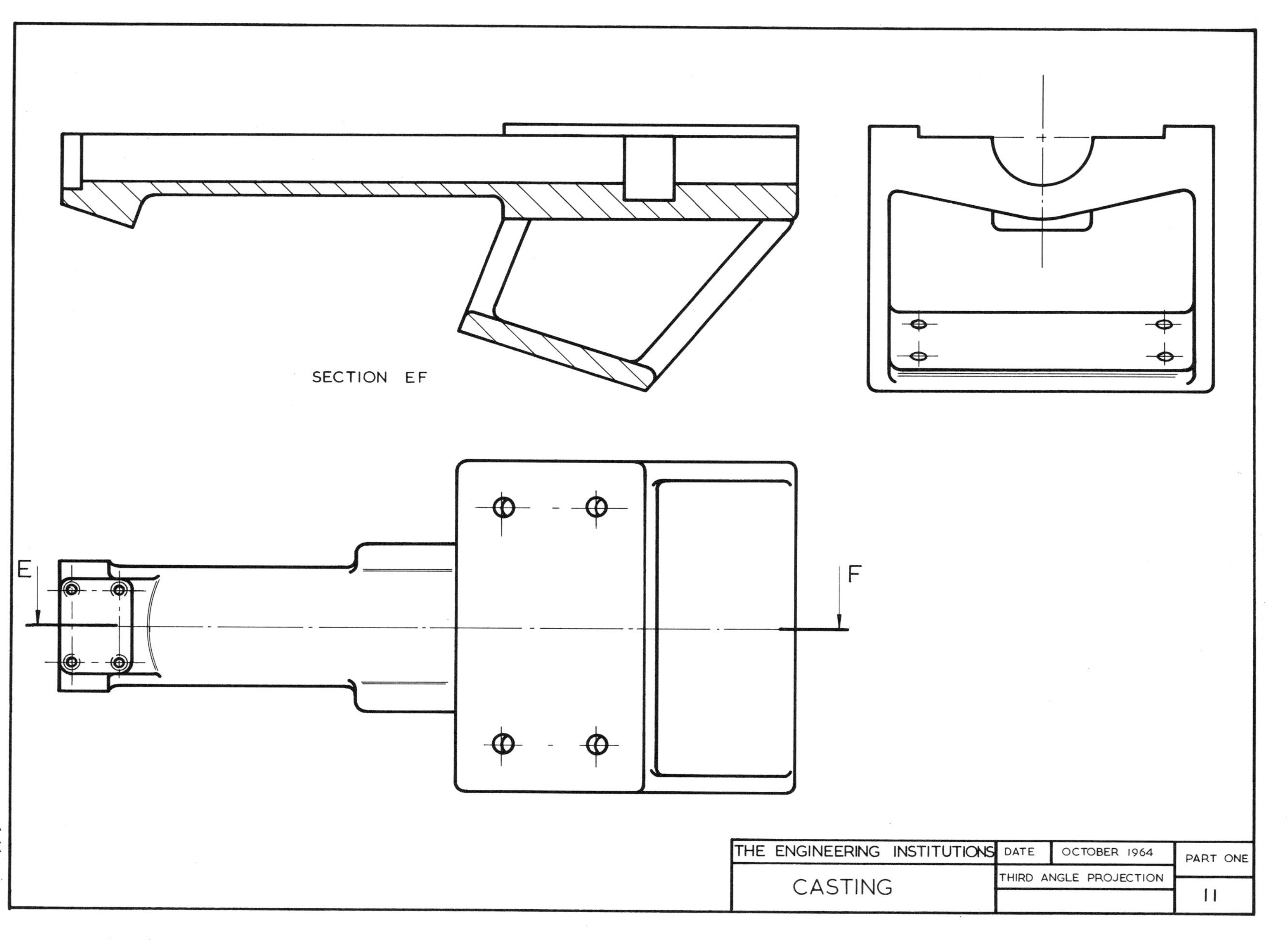
SECTION EF
E
F
THE ENGINEERING INSTITUTIONS
DATE
OCTOBER 1964
PART ONE
CASTING
THIRD ANGLE PROJECTION
11

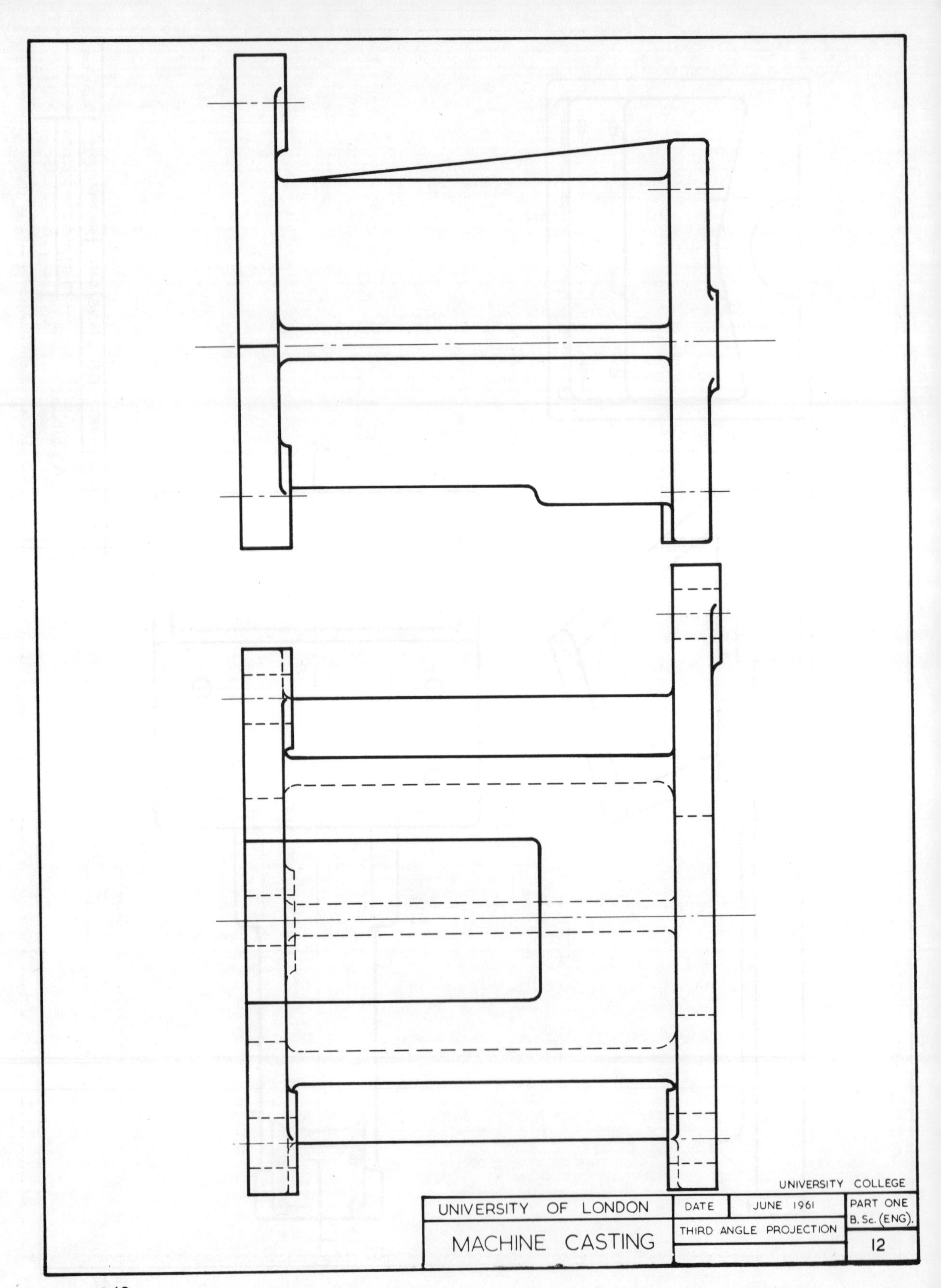
UNIVERSITY COLLEGE
UNIVERSITY OF LONDON
DATE
JUNE 1961
PART ONE
B. Sc. (ENG).
MACHINE CASTING
THIRD ANGLE PROJECTION
12

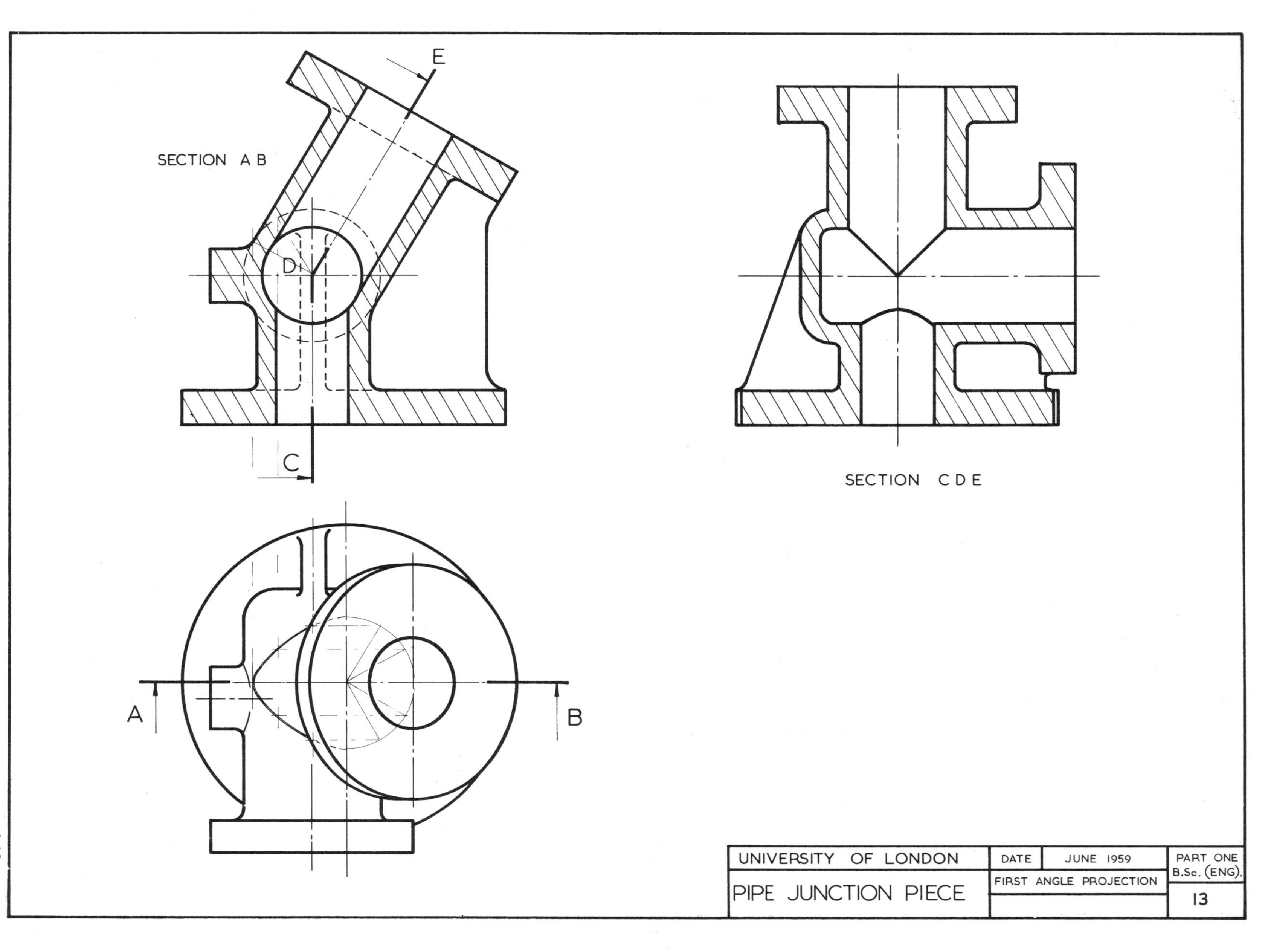
SECTION A B
E
D
C
SECTION C D E
A
B
UNIVERSITY OF LONDON
DATE
JUNE 1959
PART ONE
B.Sc. (ENG).
FIRST ANGLE PROJECTION
PIPE JUNCTION PIECE
13

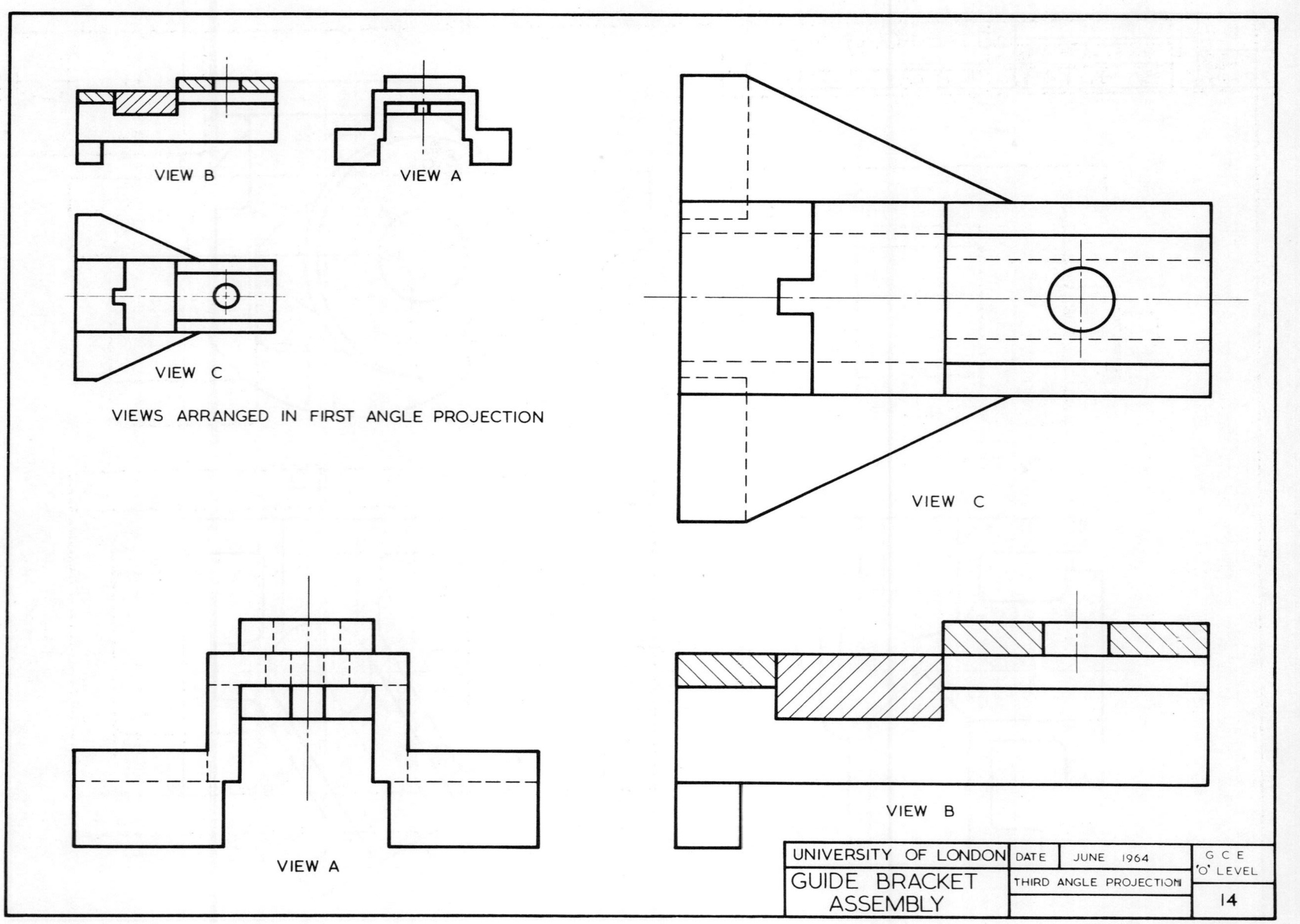

VIEW B
VIEW A
VIEW C
VIEWS ARRANGED IN FIRST ANGLE PROJECTION
VIEW C
VIEW A
VIEW B
UNIVERSITY OF LONDON
DATE
JUNE 1964
G C E
'O' LEVEL
GUIDE BRACKET
ASSEMBLY
THIRD ANGLE PROJECTION
14

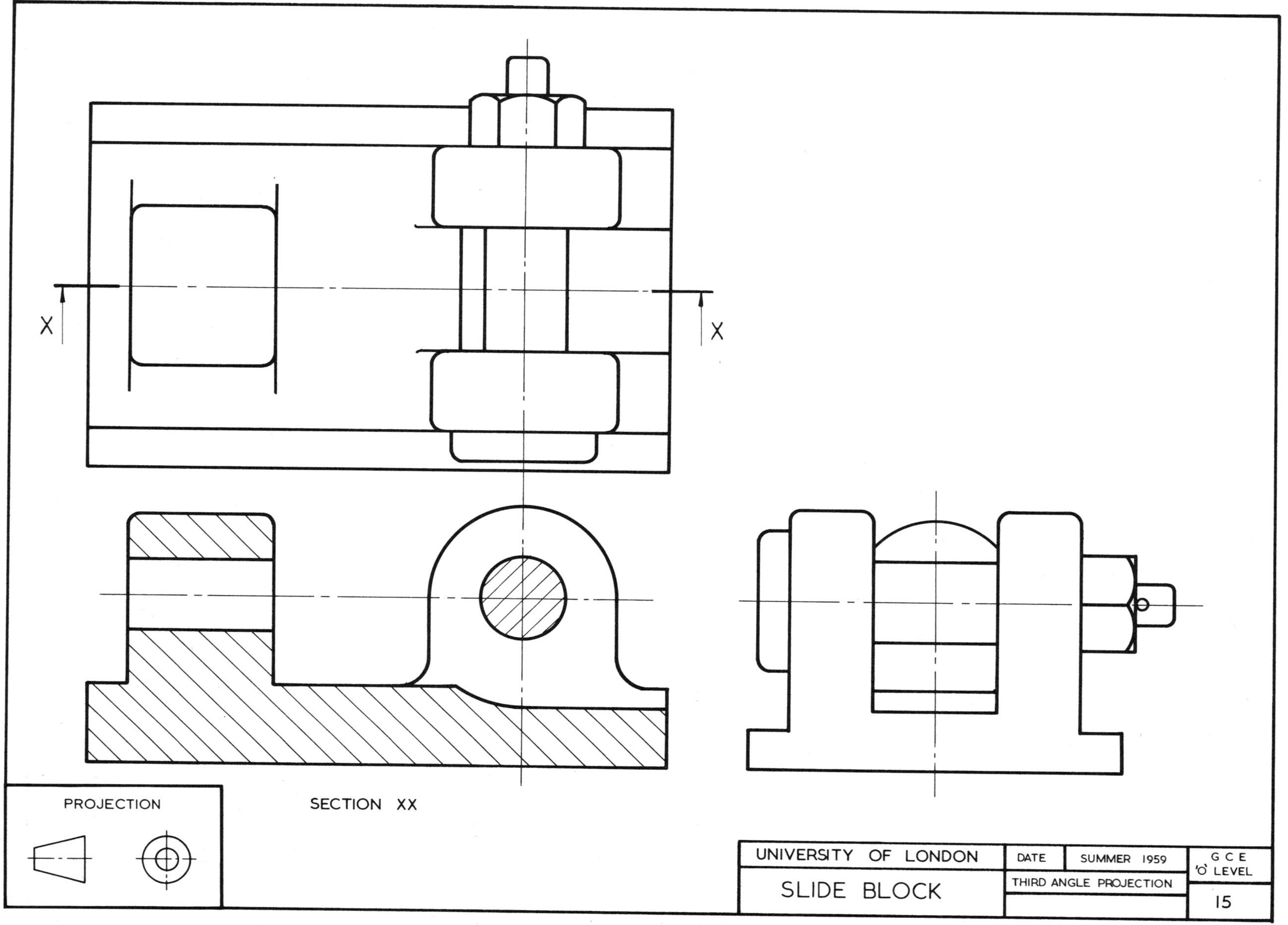
X
X
PROJECTION
SECTION XX
UNIVERSITY OF LONDON
SLIDE BLOCK
DATE
SUMMER 1959
THIRD ANGLE PROJECTION
G C E
'O' LEVEL
15

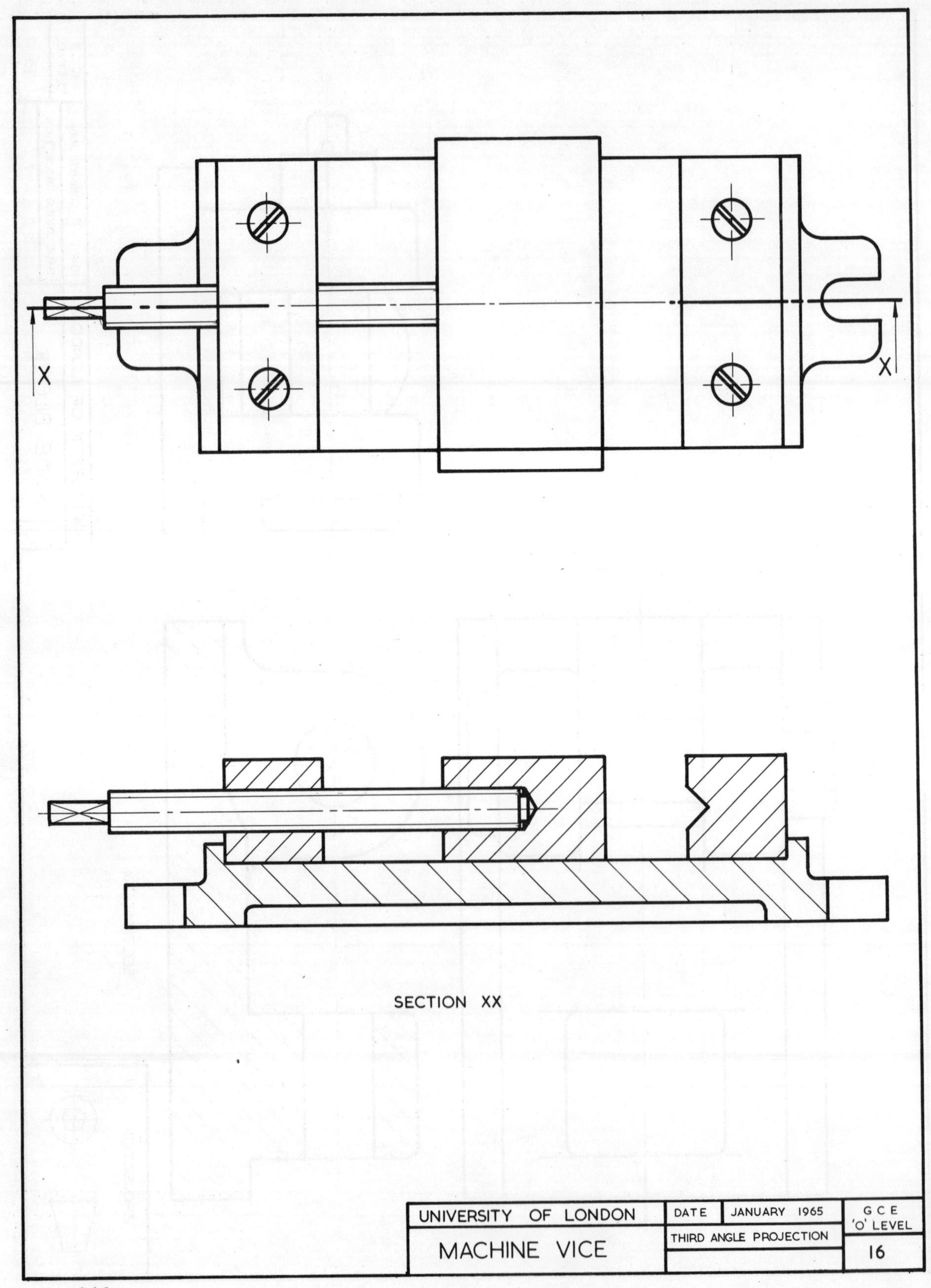
X
X
SECTION XX
UNIVERSITY OF LONDON
MACHINE VICE
DATE
JANUARY 1965
THIRD ANGLE PROJECTION
G C E
'O' LEVEL
16

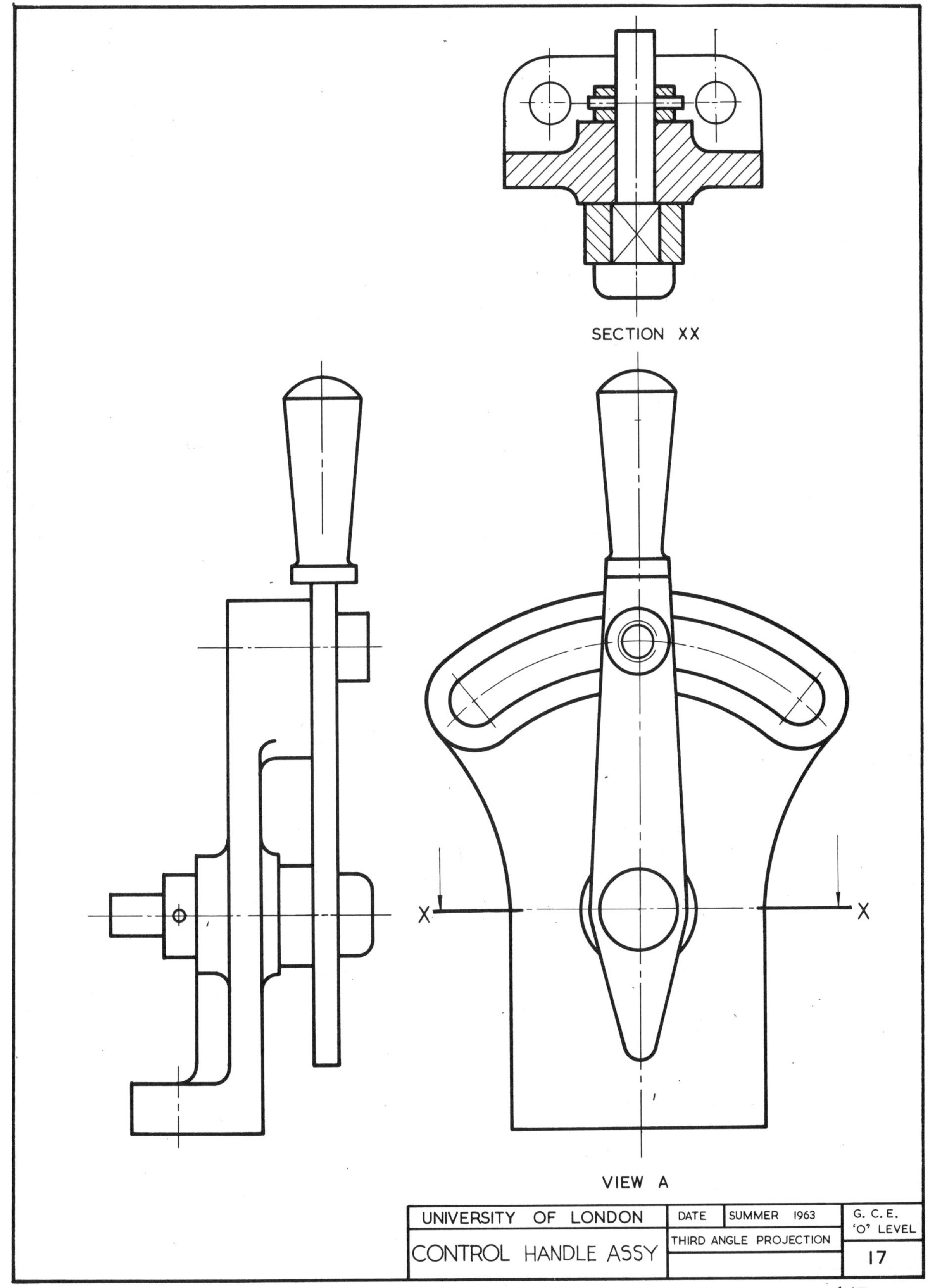
SECTION XX
X
X
VIEW A
UNIVERSITY OF LONDON
DATE
SUMMER 1963
G. C. E. 'O' LEVEL
THIRD ANGLE PROJECTION
CONTROL HANDLE ASSY
17

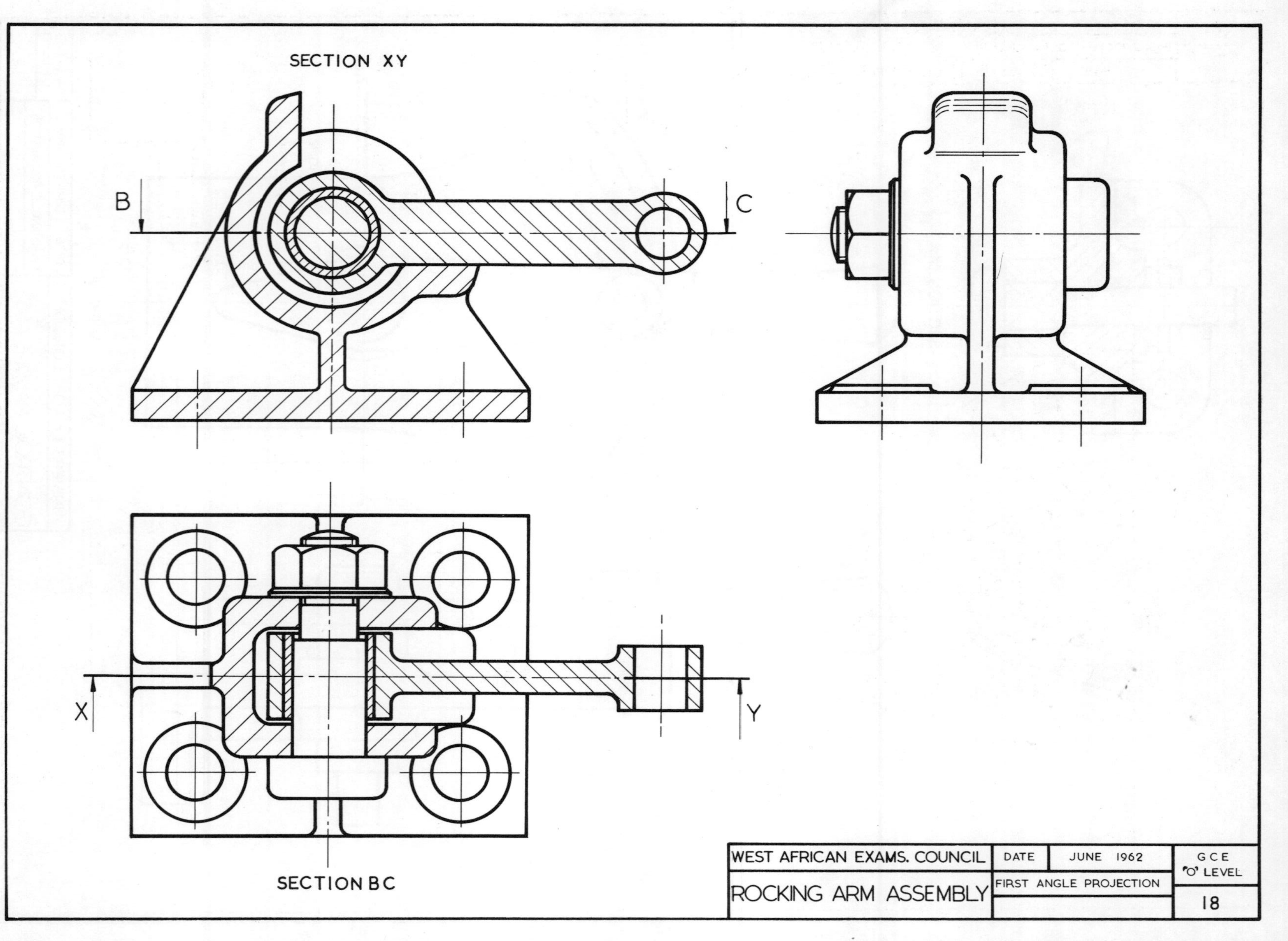
SECTION XY
B
C
X
Y
SECTION BC
WEST AFRICAN EXAMS. COUNCIL
DATE
JUNE 1962
G C E
'O' LEVEL
ROCKING ARM ASSEMBLY
FIRST ANGLE PROJECTION
18

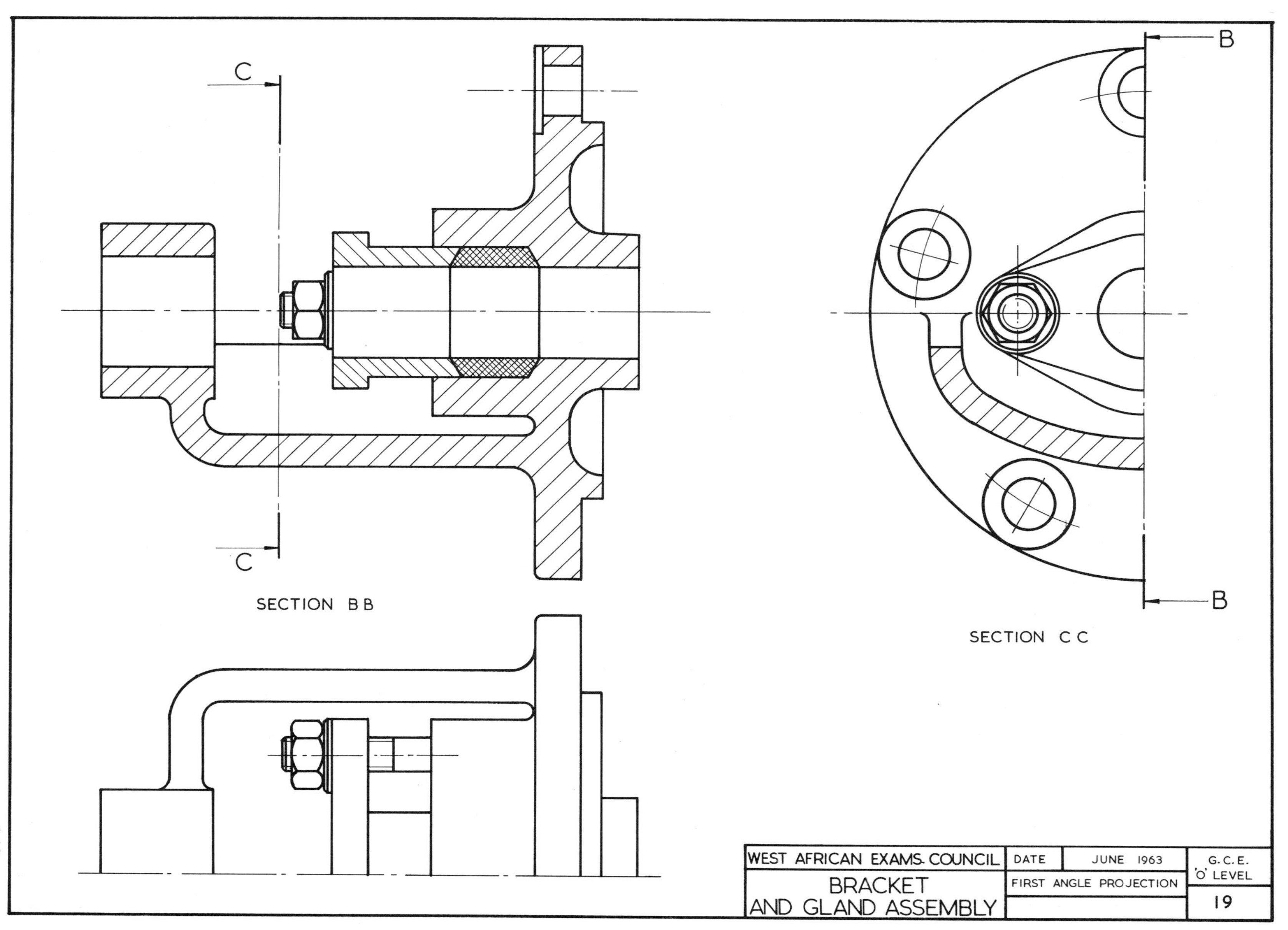
C
C
SECTION B B
B
B
SECTION C C
WEST AFRICAN EXAMS. COUNCIL
BRACKET
AND GLAND ASSEMBLY
DATE
JUNE 1963
FIRST ANGLE PROJECTION
G.C.E.
'O' LEVEL
19

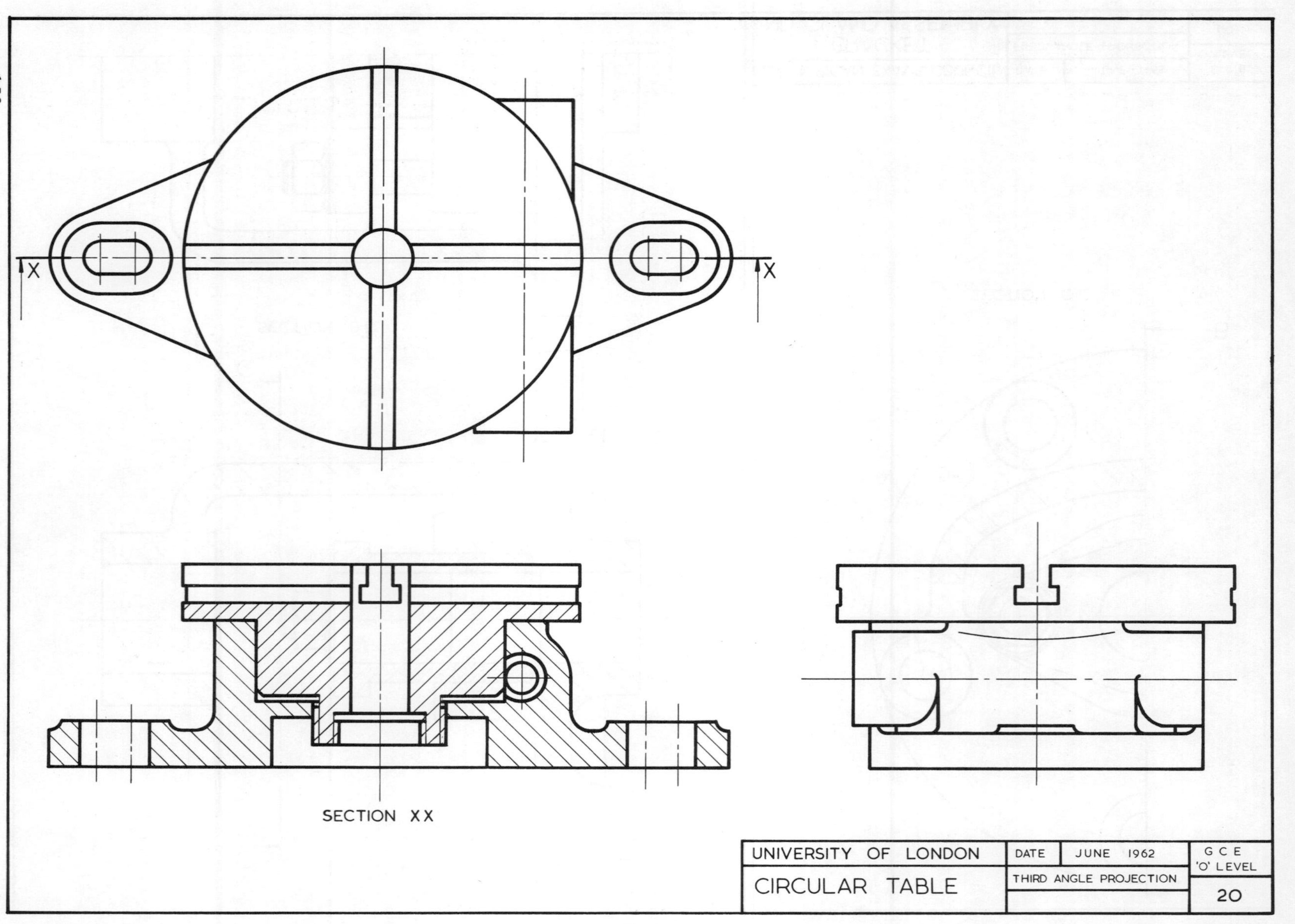

X
X
SECTION XX
UNIVERSITY OF LONDON
CIRCULAR TABLE
DATE
JUNE 1962
THIRD ANGLE PROJECTION
G C E
'O' LEVEL
20

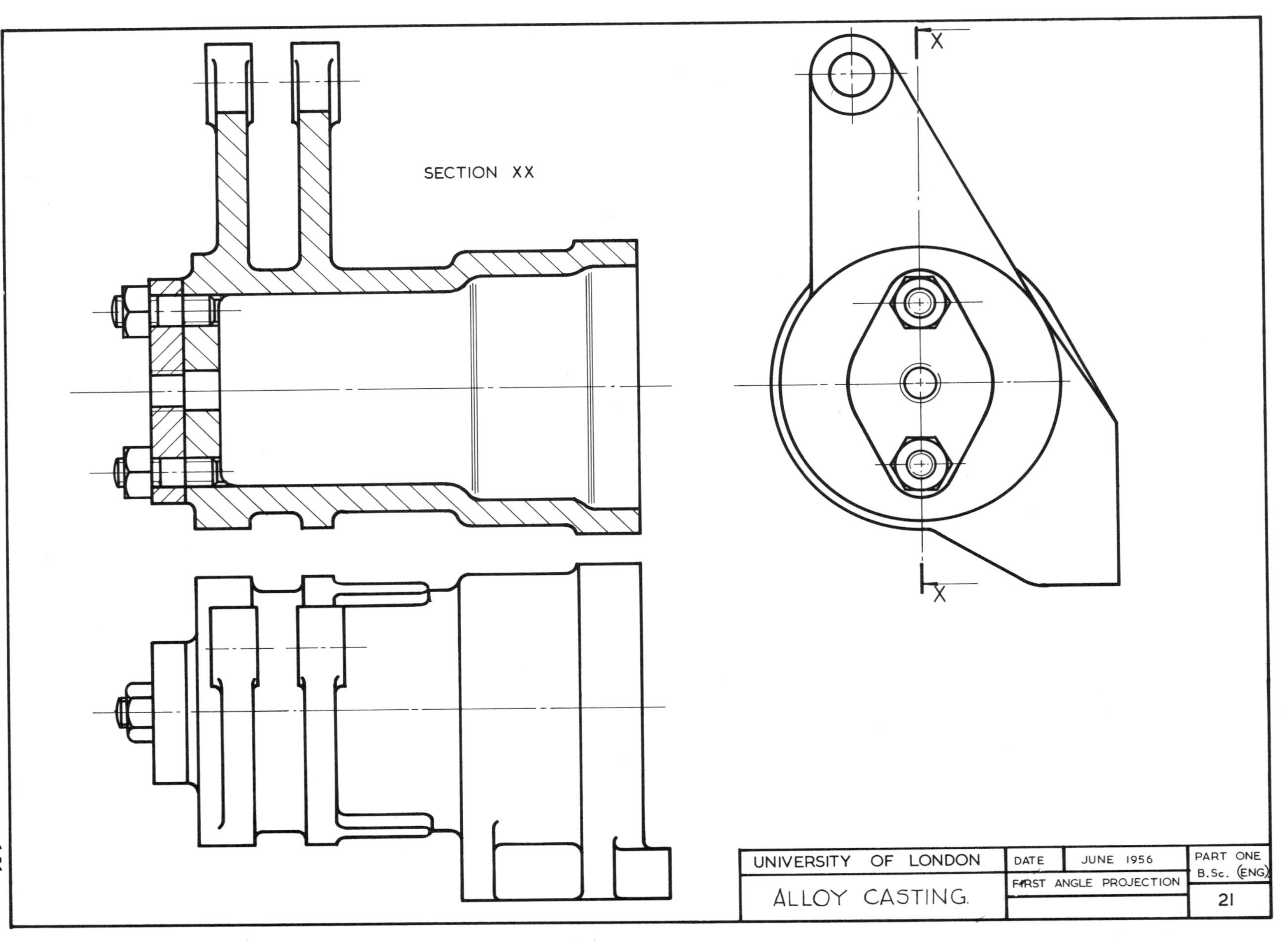
SECTION XX
X
X
UNIVERSITY OF LONDON
ALLOY CASTING.
DATE
JUNE 1956
FIRST ANGLE PROJECTION
PART ONE
B.Sc. (ENG)
21

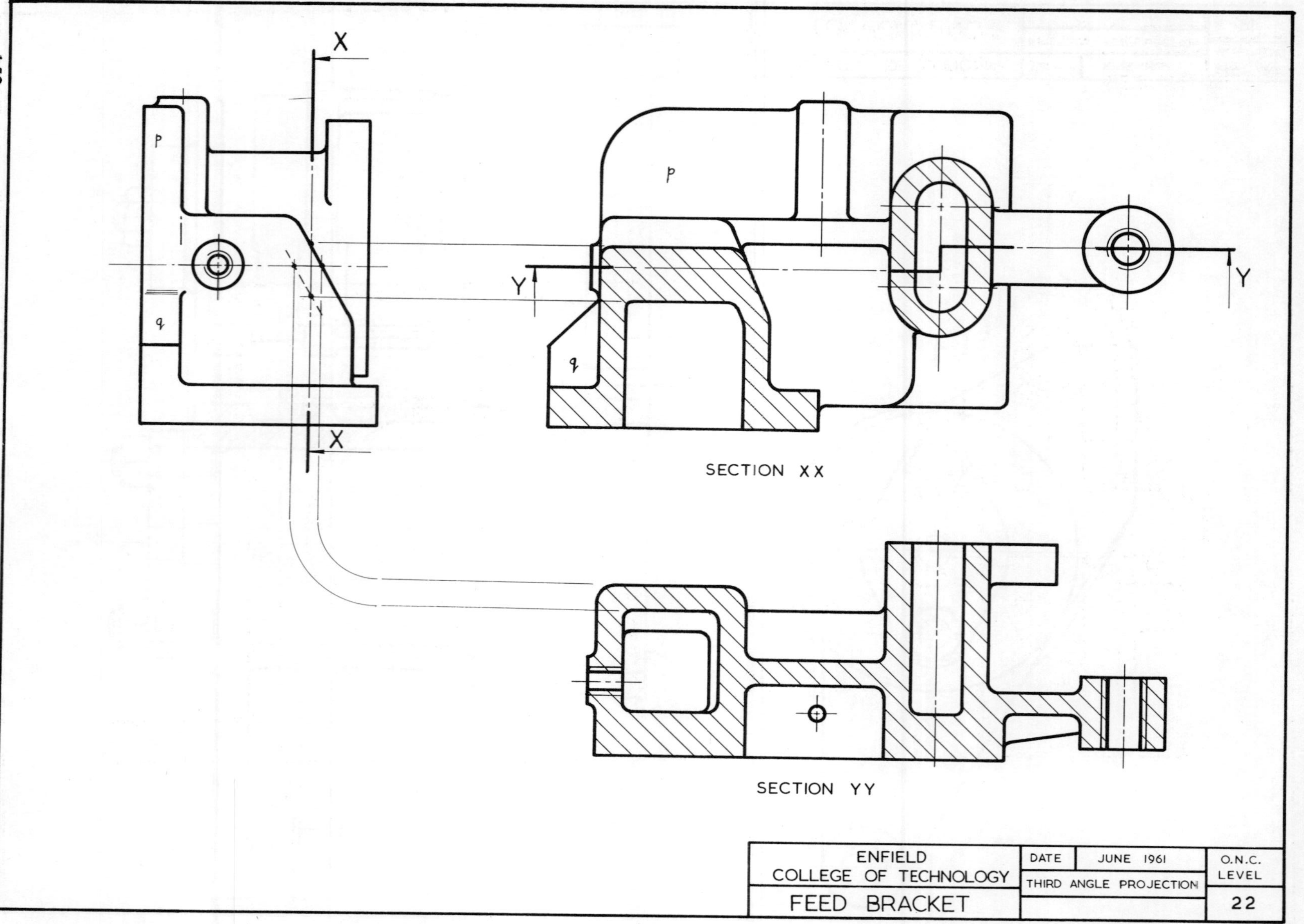
X
X
Y
Y
p
q
p
q
SECTION XX
SECTION YY
ENFIELD
COLLEGE OF TECHNOLOGY
FEED BRACKET
DATE
JUNE 1961
THIRD ANGLE PROJECTION
O.N.C.
LEVEL
22

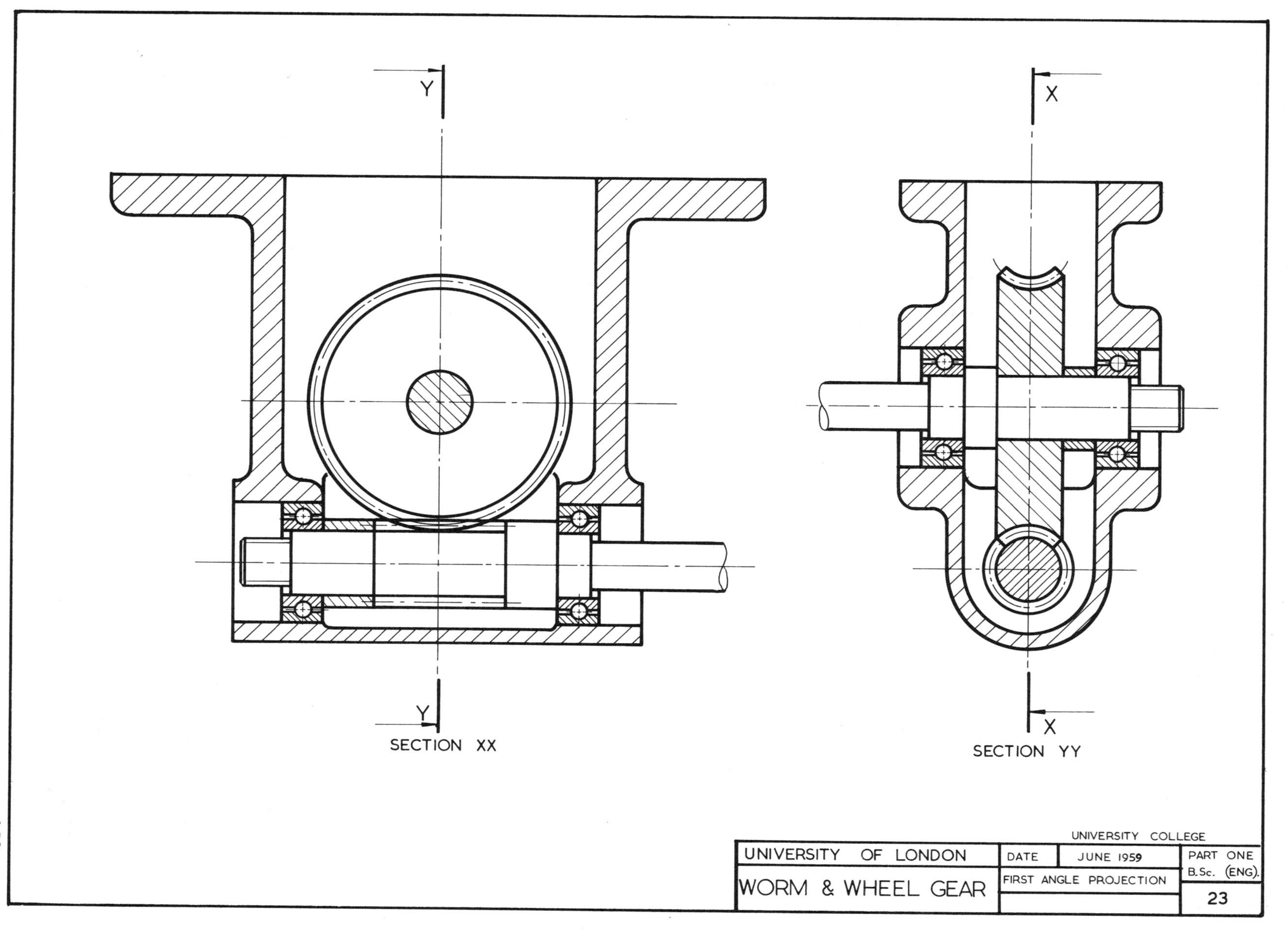
Y
X
Y
SECTION XX
X
SECTION YY
UNIVERSITY COLLEGE
UNIVERSITY OF LONDON
DATE
JUNE 1959
PART ONE
B.Sc. (ENG).
WORM & WHEEL GEAR
FIRST ANGLE PROJECTION
23

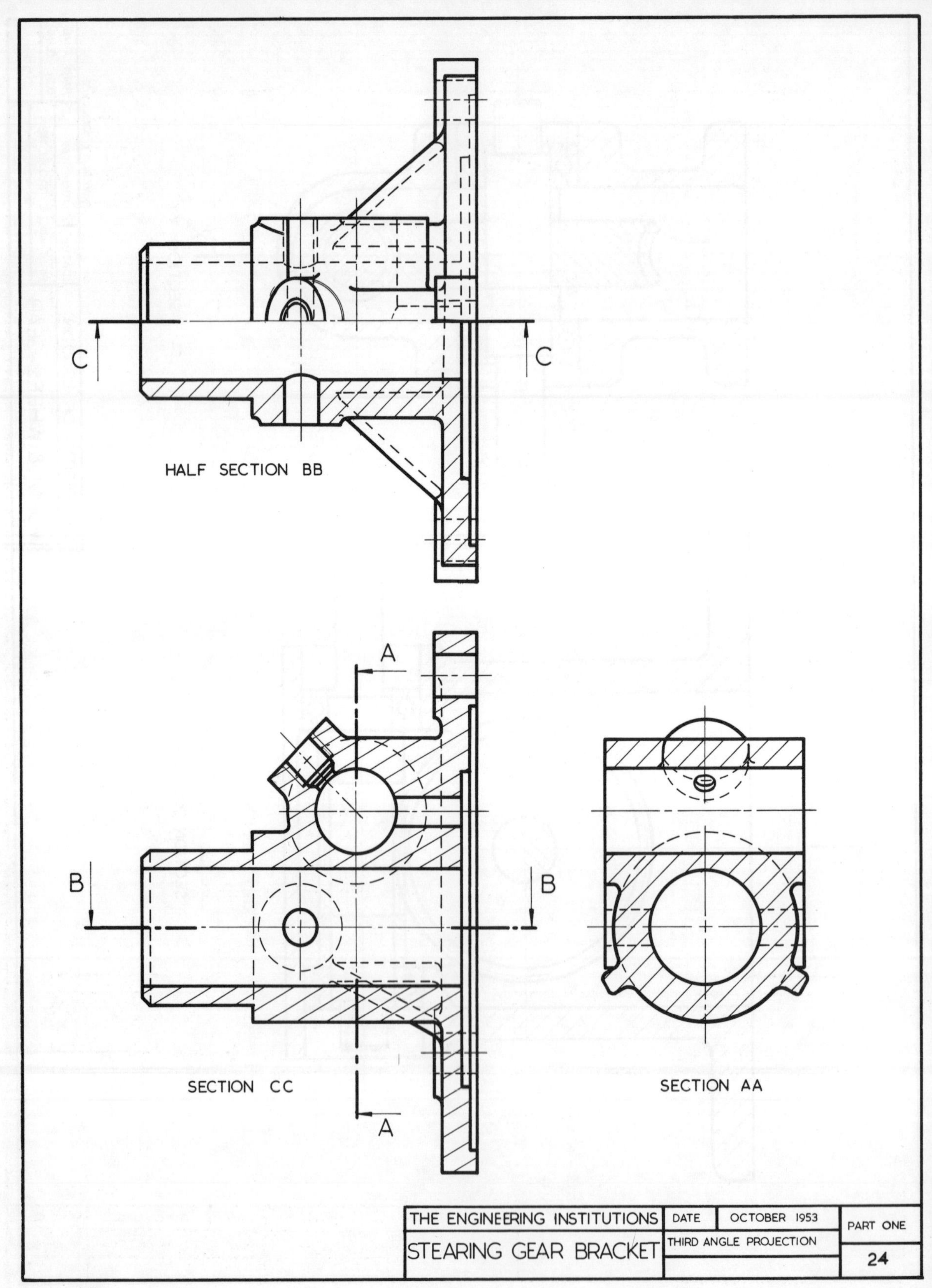
HALF SECTION BB
C
C
A
B
B
A
SECTION CC
SECTION AA
THE ENGINEERING INSTITUTIONS
DATE
OCTOBER 1953
PART ONE
STEARING GEAR BRACKET
THIRD ANGLE PROJECTION
24

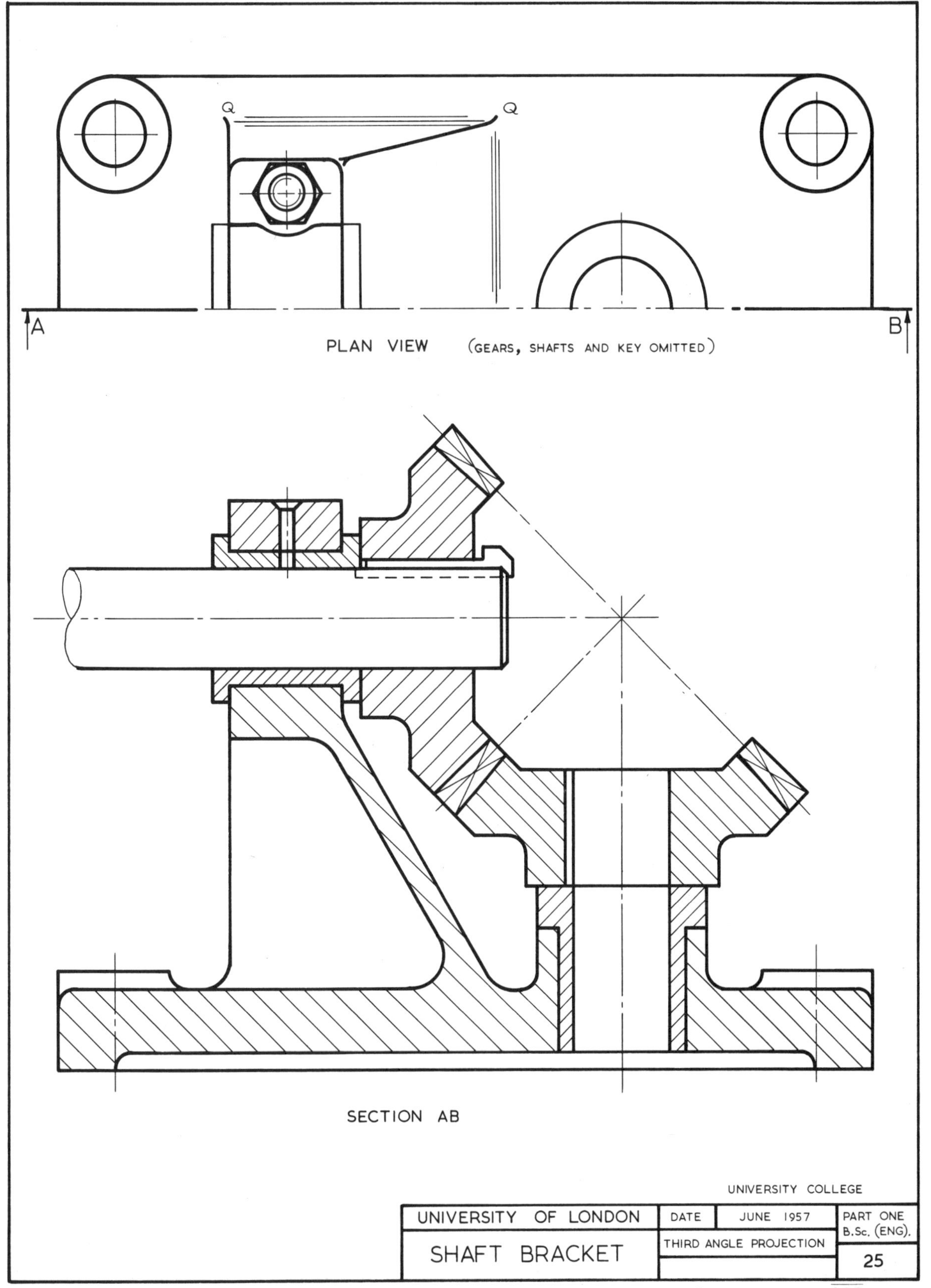
Q
Q
A
B
PLAN VIEW (GEARS, SHAFTS AND KEY OMITTED)
SECTION AB
UNIVERSITY COLLEGE
UNIVERSITY OF LONDON
DATE
JUNE 1957
PART ONE
B.Sc. (ENG).
SHAFT BRACKET
THIRD ANGLE PROJECTION
25

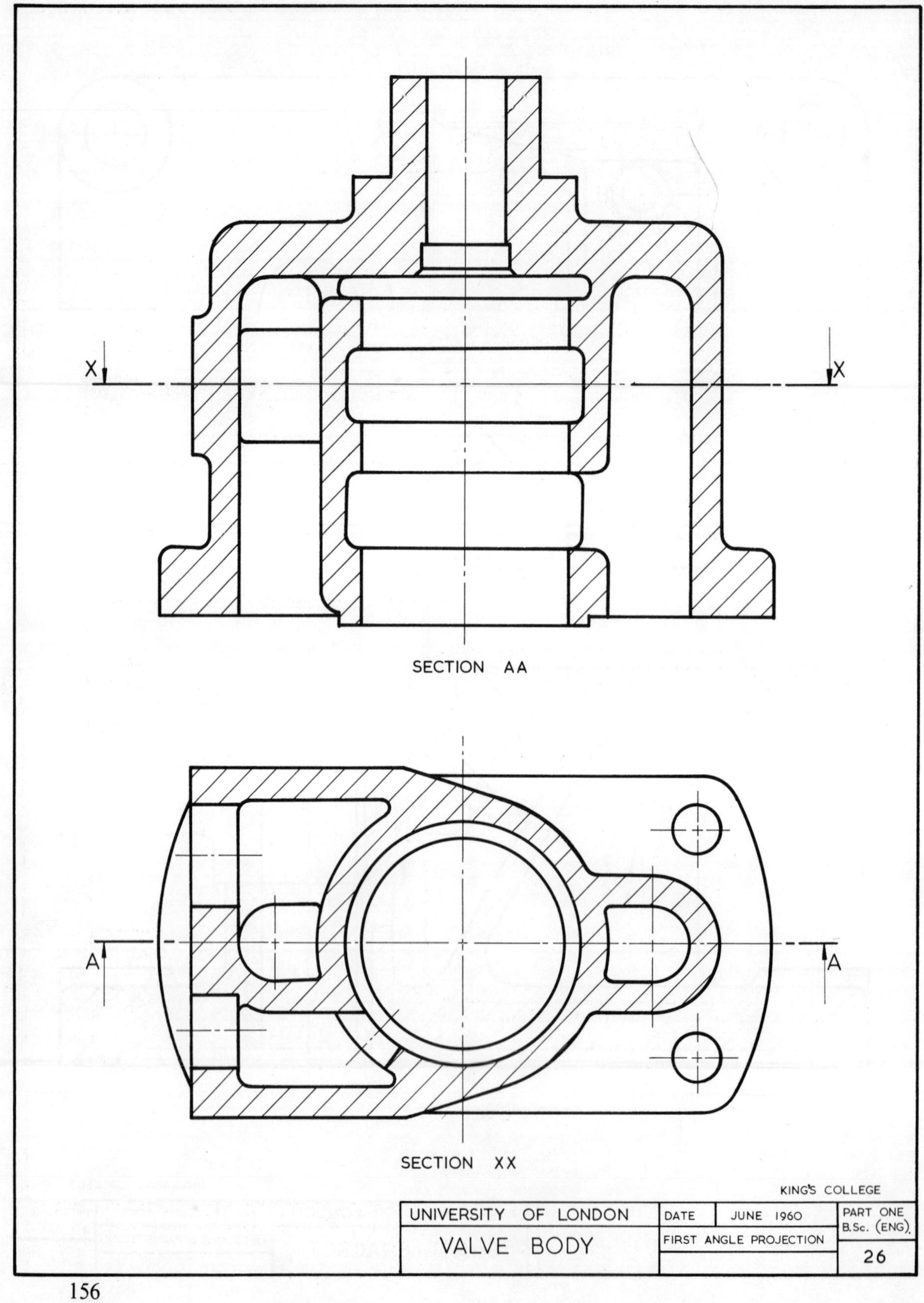
X
X
SECTION AA
A
A
SECTION XX
KING'S COLLEGE
UNIVERSITY OF LONDON
DATE
JUNE 1960
PART ONE B.Sc. (ENG).
VALVE BODY
FIRST ANGLE PROJECTION
26

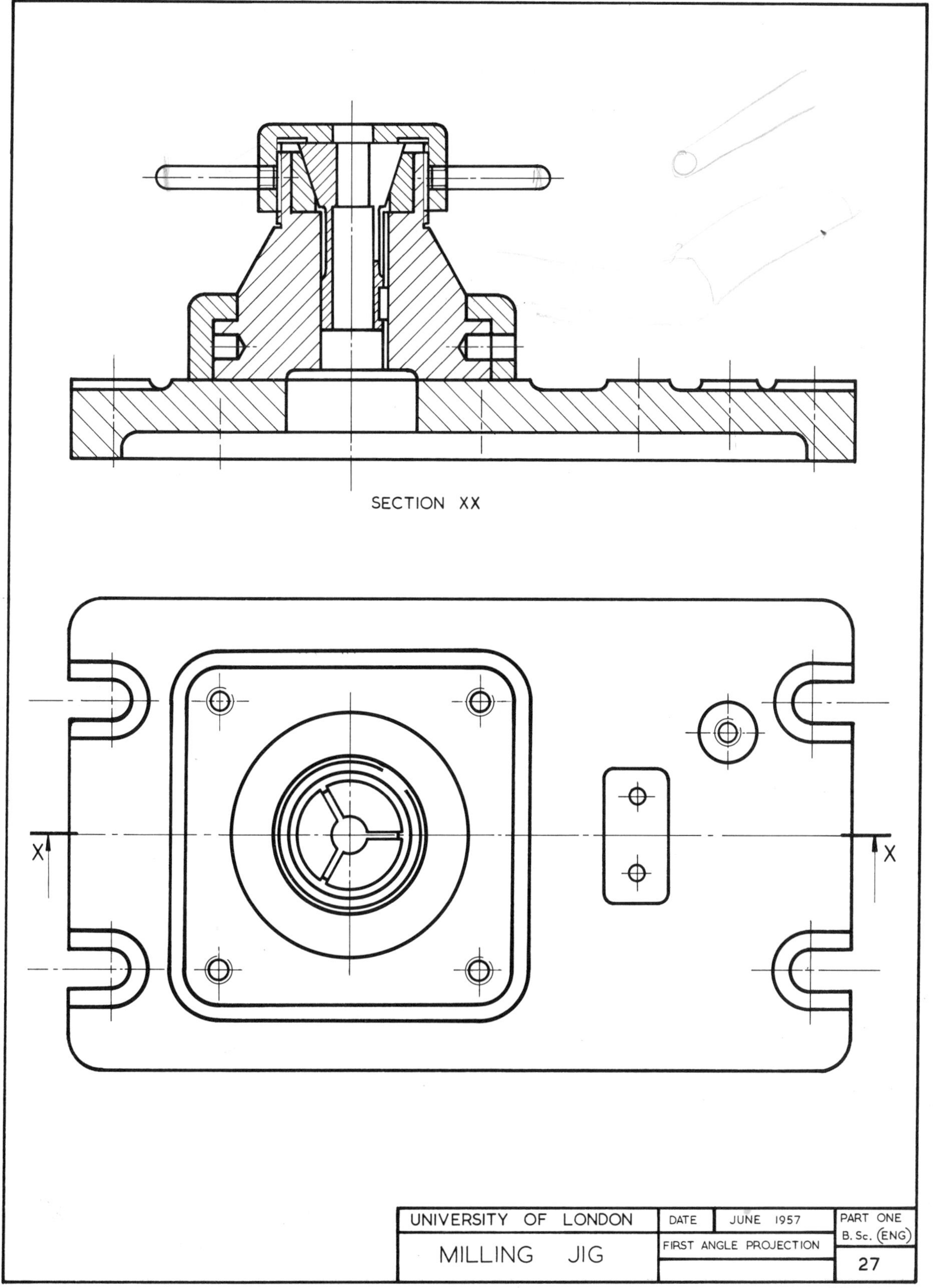
SECTION XX
X
X
UNIVERSITY OF LONDON
DATE
JUNE 1957
PART ONE
B. Sc. (ENG)
MILLING JIG
FIRST ANGLE PROJECTION
27

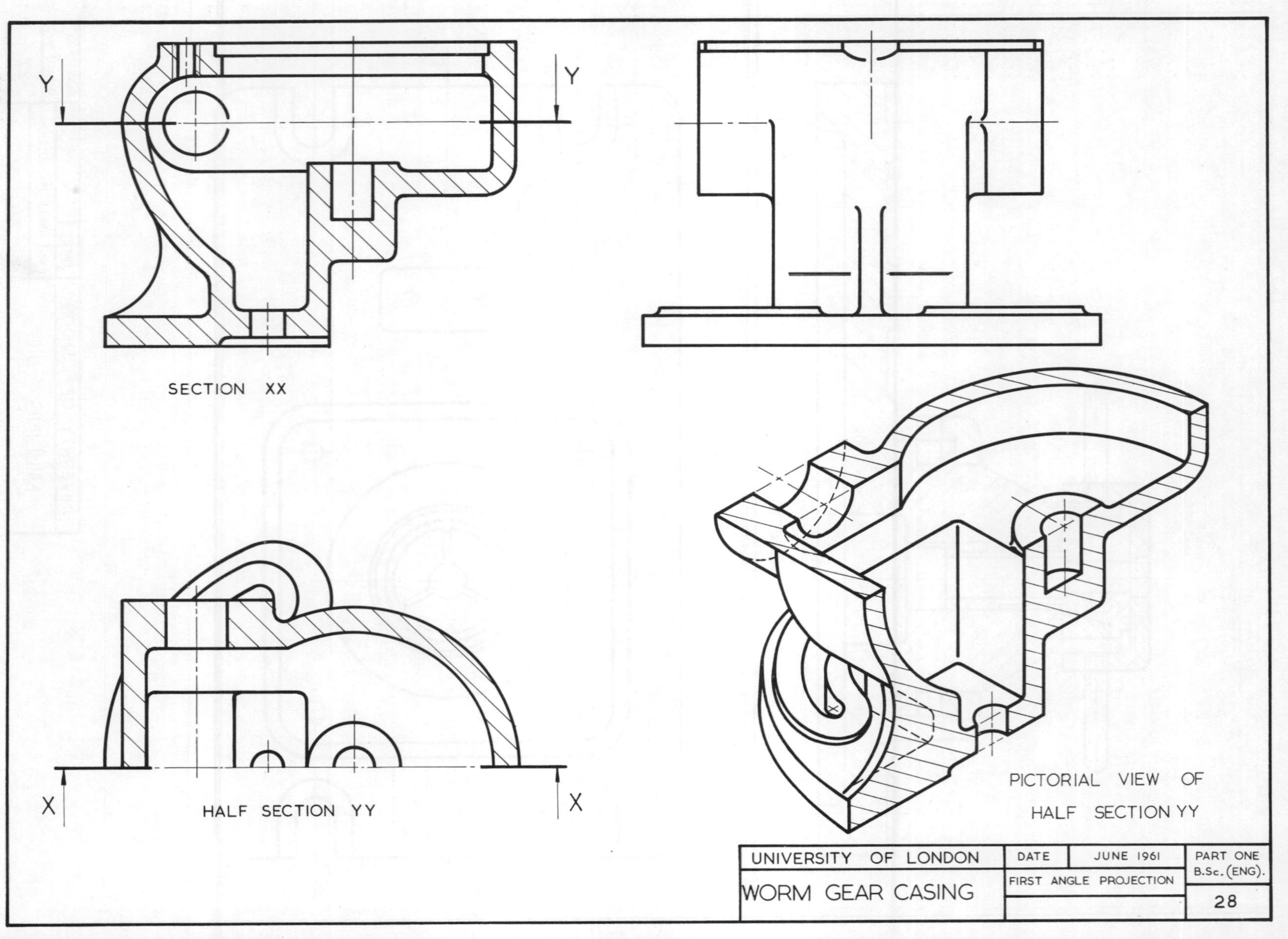
Y
Y
SECTION XX
X
HALF SECTION YY
X
PICTORIAL VIEW OF
HALF SECTION YY
UNIVERSITY OF LONDON
WORM GEAR CASING
DATE
JUNE 1961
FIRST ANGLE PROJECTION
PART ONE
B.Sc. (ENG).
28

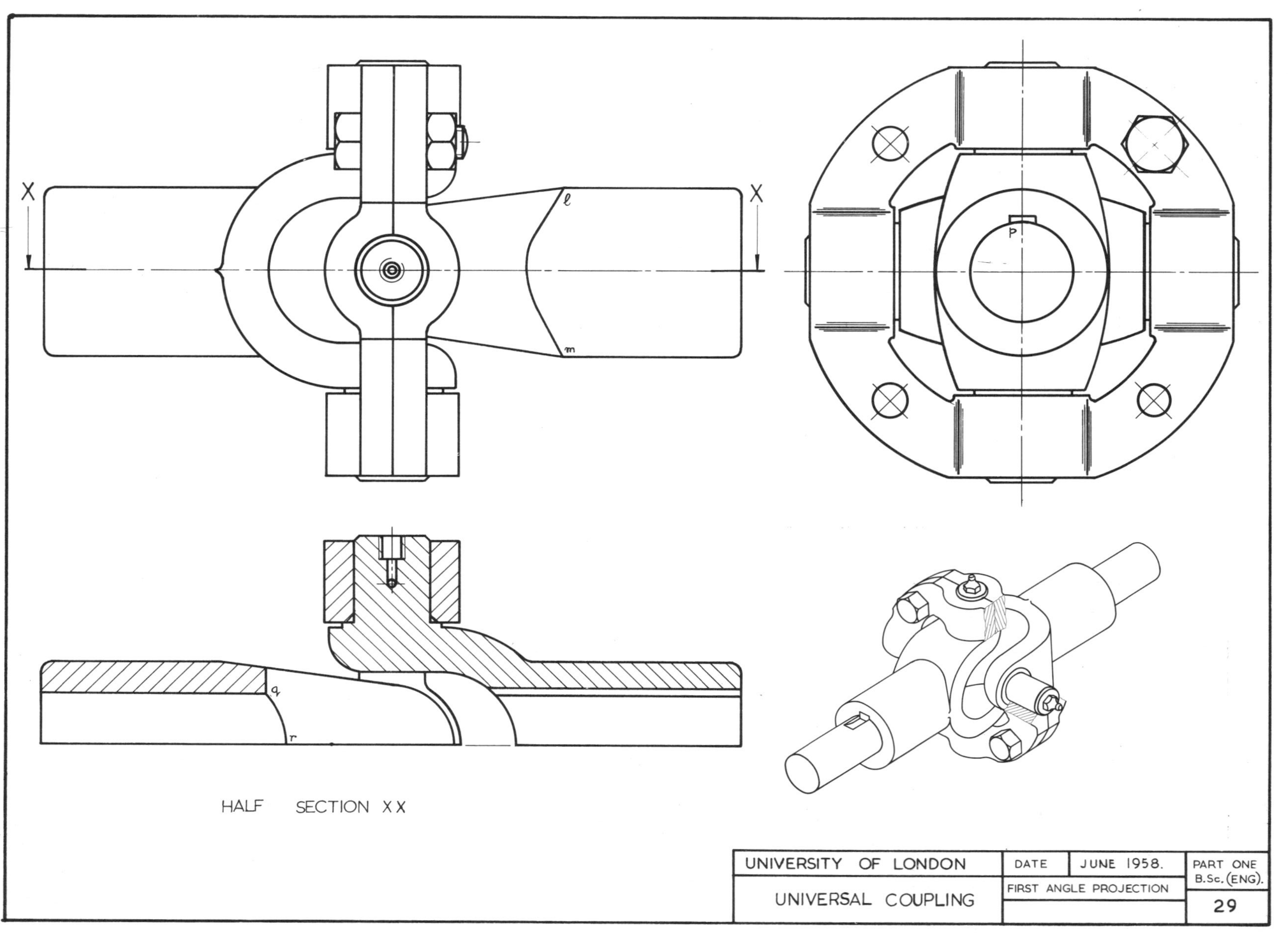
X
X
l
m
P
HALF SECTION XX
q
r
UNIVERSITY OF LONDON
DATE
JUNE 1958.
PART ONE
B.Sc. (ENG).
UNIVERSAL COUPLING
FIRST ANGLE PROJECTION
29

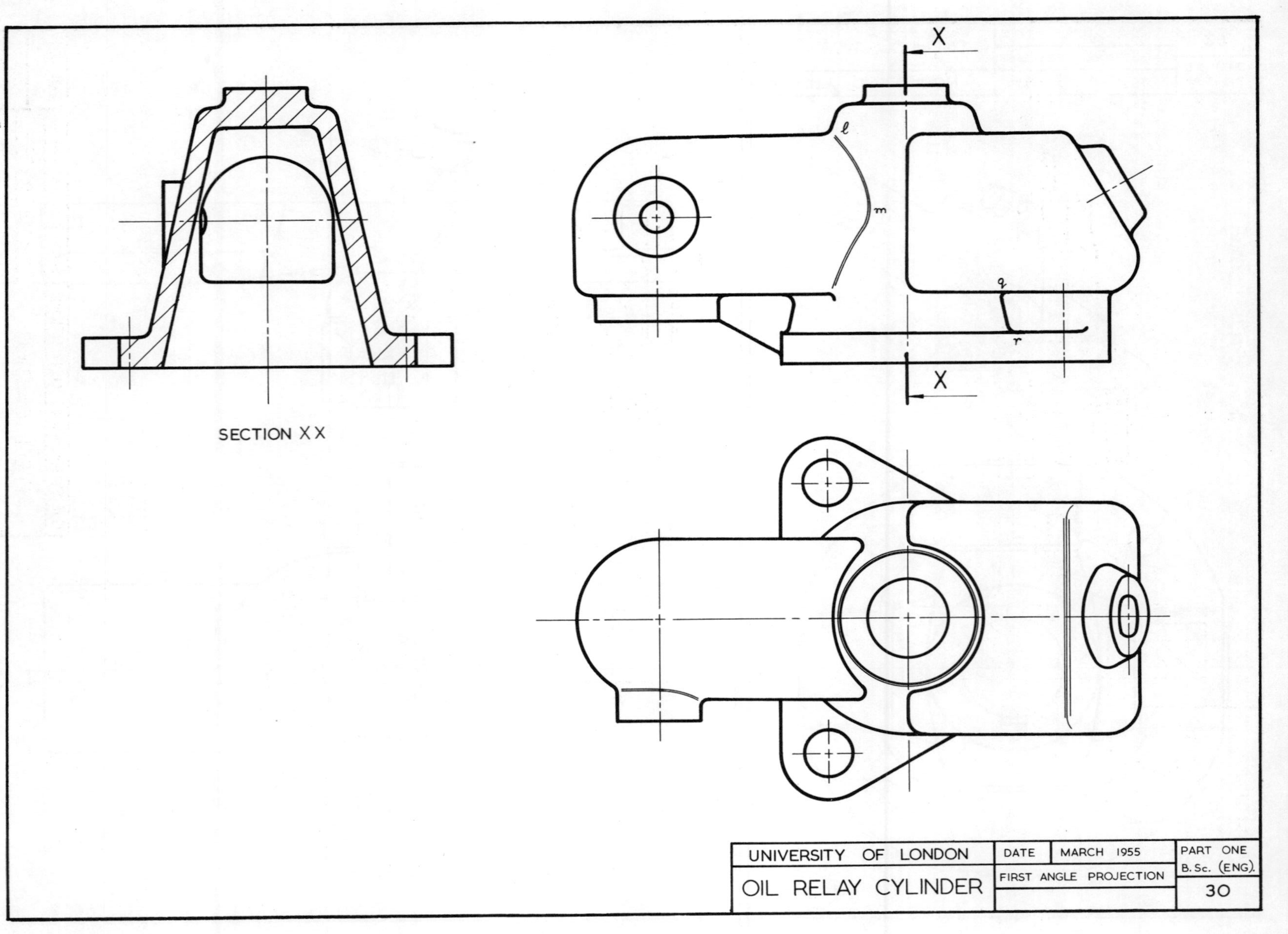

SECTION X X
X
X
UNIVERSITY OF LONDON
DATE
MARCH 1955
PART ONE
B. Sc. (ENG.)
OIL RELAY CYLINDER
FIRST ANGLE PROJECTION
30

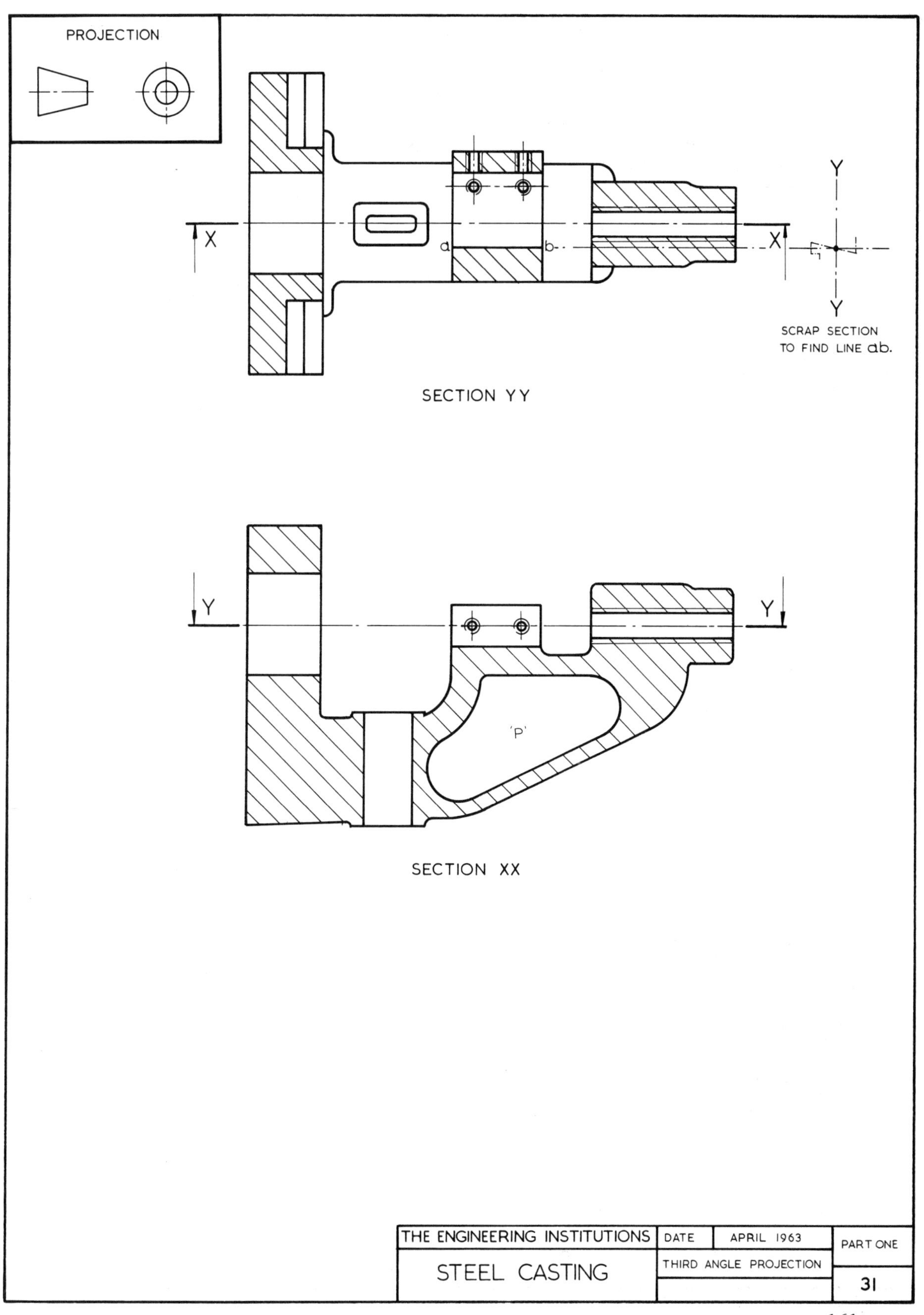
PROJECTION
X
X
a
b
Y
Y
SCRAP SECTION
TO FIND LINE ab.
SECTION YY
Y
Y
'P'
SECTION XX
THE ENGINEERING INSTITUTIONS
DATE
APRIL 1963
PART ONE
STEEL CASTING
THIRD ANGLE PROJECTION
31

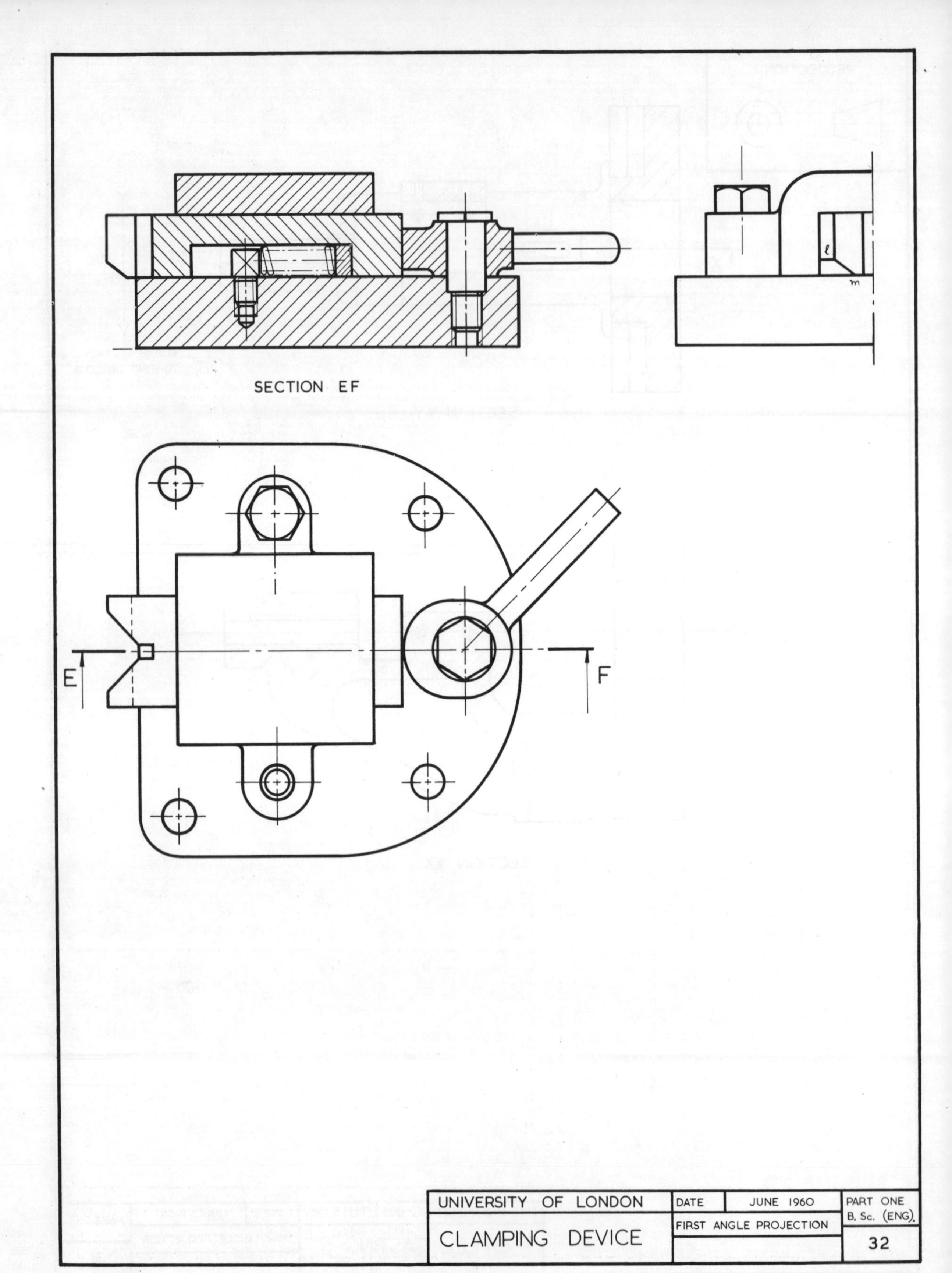
SECTION EF
E
F
l
m
X
UNIVERSITY OF LONDON
DATE
JUNE 1960
PART ONE
B. Sc. (ENG).
FIRST ANGLE PROJECTION
CLAMPING DEVICE
32

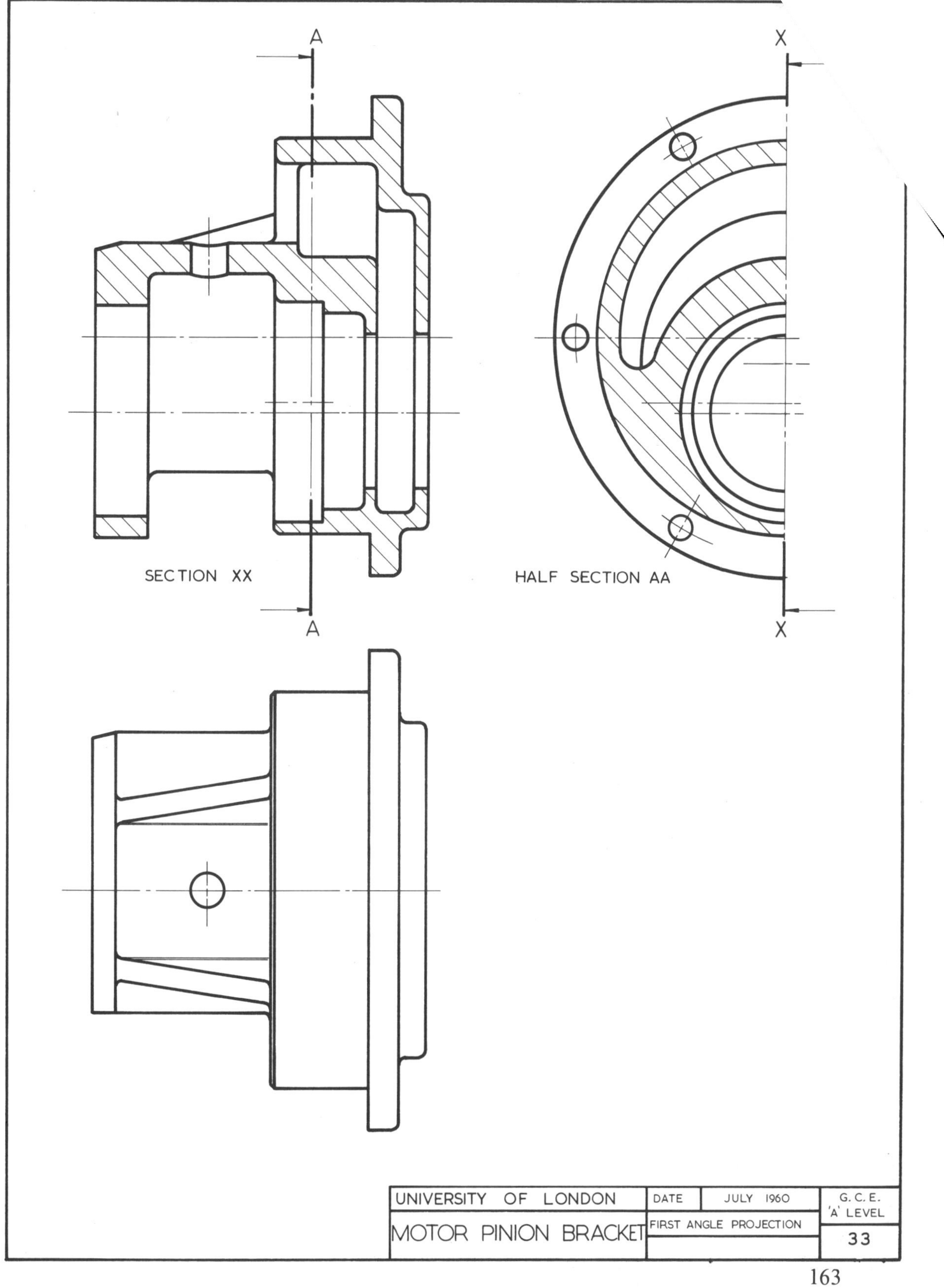
A
X
SECTION XX
HALF SECTION AA
A
X
UNIVERSITY OF LONDON
DATE
JULY 1960
G. C. E.
'A' LEVEL
MOTOR PINION BRACKET
FIRST ANGLE PROJECTION
33

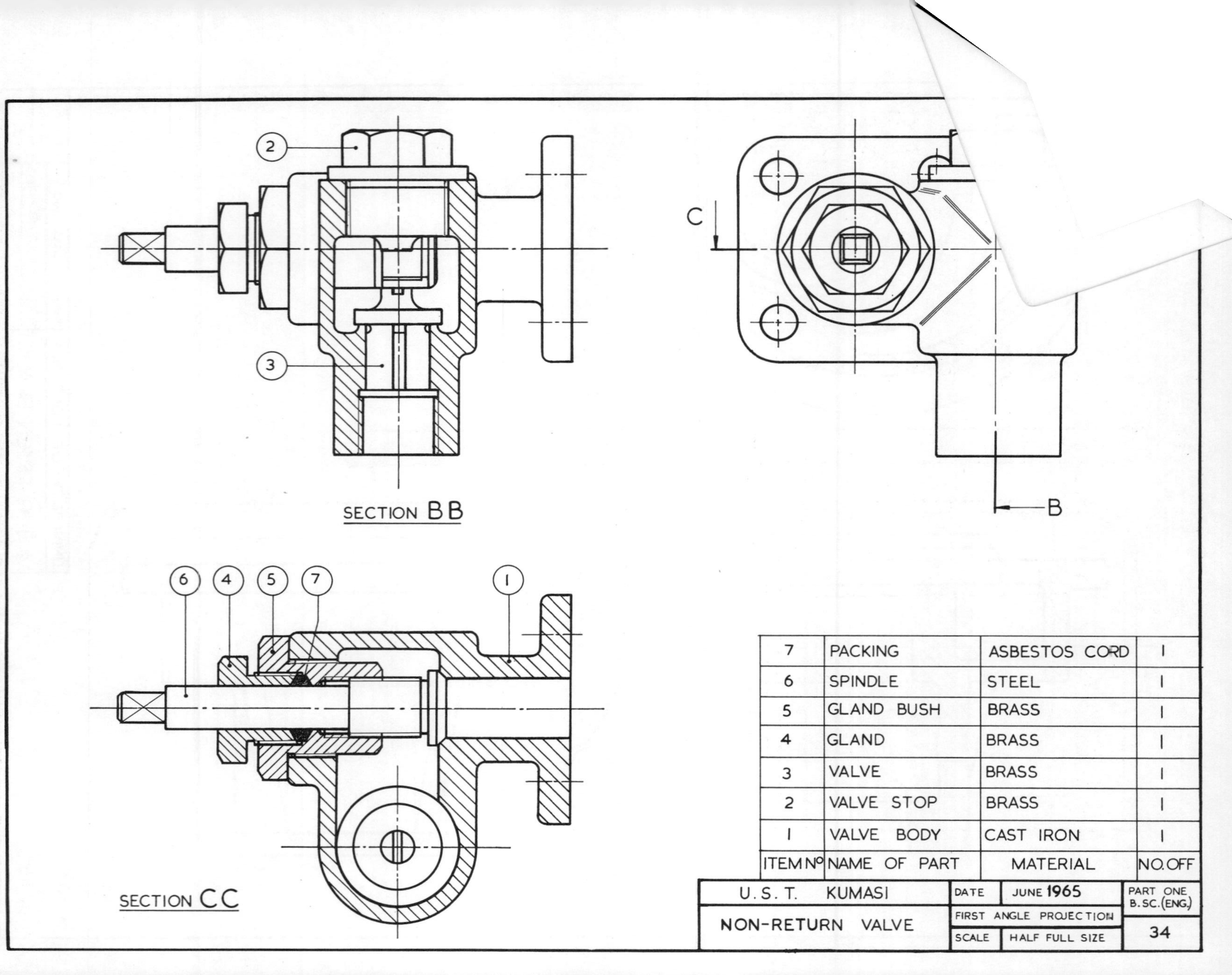

ITEM Nº	NAME OF PART	MATERIAL	NO. OFF
7	PACKING	ASBESTOS CORD	1
6	SPINDLE	STEEL	1
5	GLAND BUSH	BRASS	1
4	GLAND	BRASS	1
3	VALVE	BRASS	1
2	VALVE STOP	BRASS	1
1	VALVE BODY	CAST IRON	1

U. S. T. KUMASI	DATE	JUNE 1965	PART ONE B.SC. (ENG.)
NON-RETURN VALVE	FIRST ANGLE PROJECTION		34
	SCALE	HALF FULL SIZE	

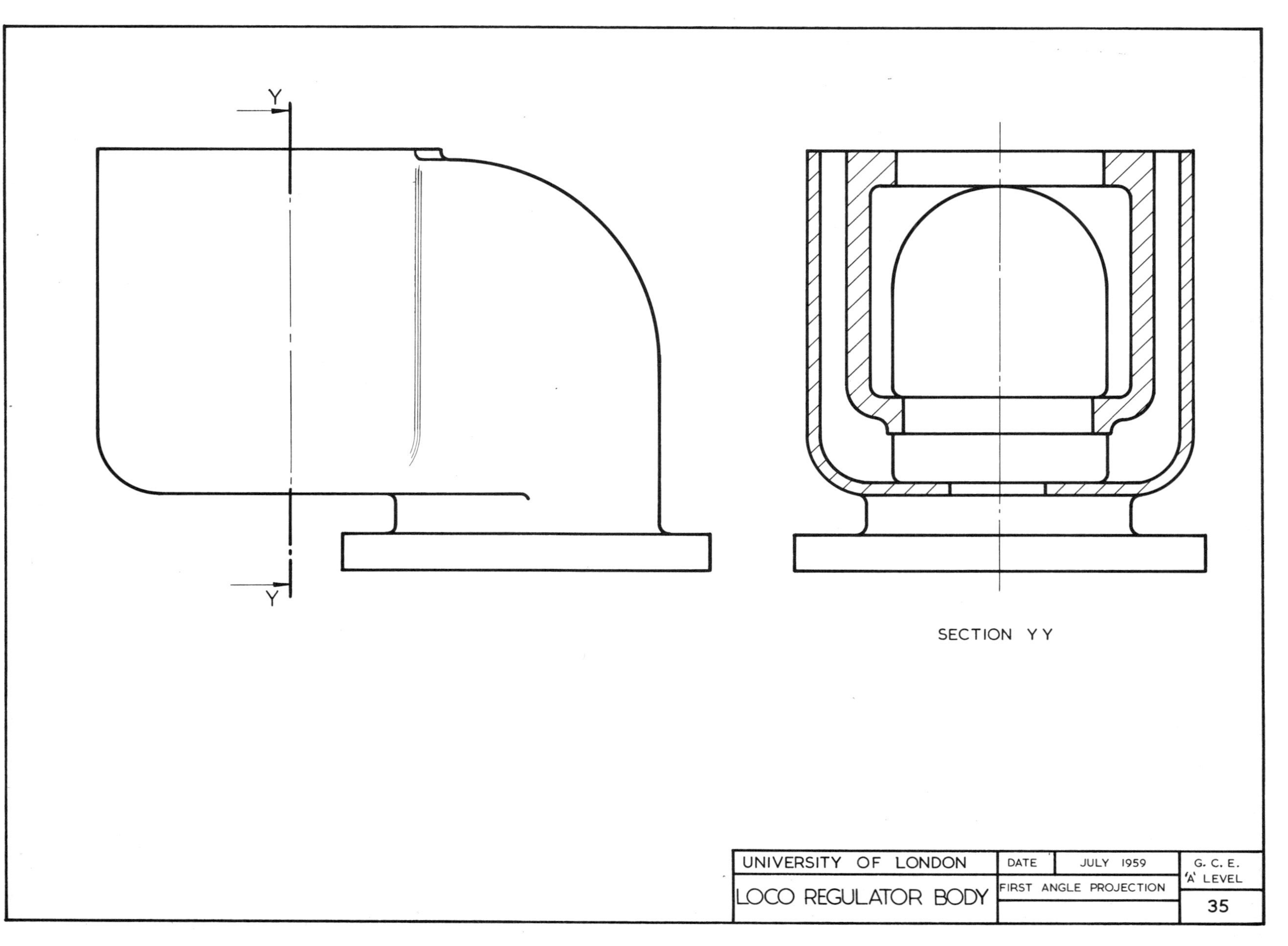
Y
Y
SECTION Y Y
UNIVERSITY OF LONDON
DATE
JULY 1959
G. C. E.
'A' LEVEL
LOCO REGULATOR BODY
FIRST ANGLE PROJECTION
35

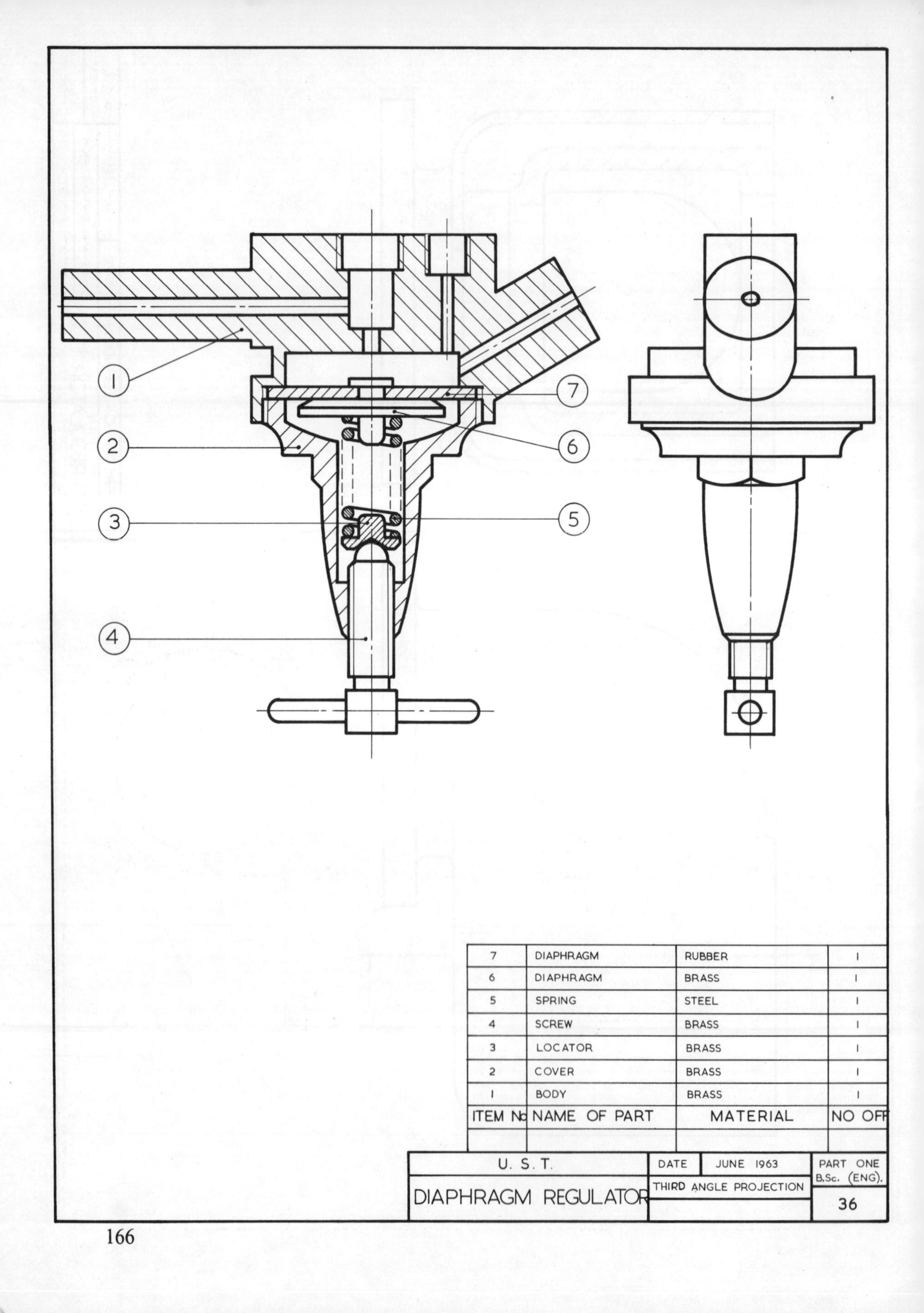

ITEM No	NAME OF PART	MATERIAL	NO OFF
7	DIAPHRAGM	RUBBER	1
6	DIAPHRAGM	BRASS	1
5	SPRING	STEEL	1
4	SCREW	BRASS	1
3	LOCATOR	BRASS	1
2	COVER	BRASS	1
1	BODY	BRASS	1

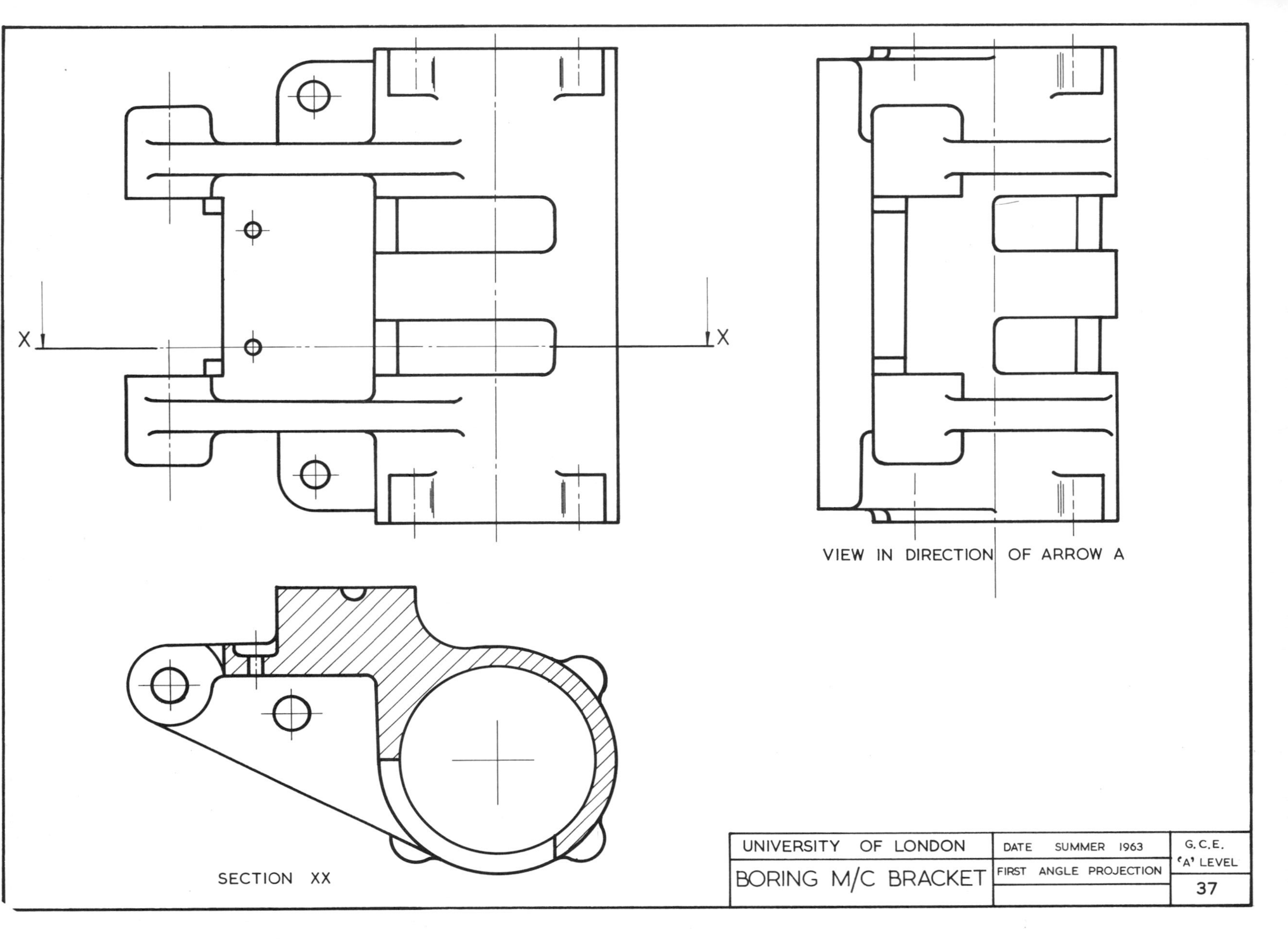

X
X
VIEW IN DIRECTION OF ARROW A
SECTION XX
UNIVERSITY OF LONDON
BORING M/C BRACKET
DATE SUMMER 1963
FIRST ANGLE PROJECTION
G.C.E.
'A' LEVEL
37

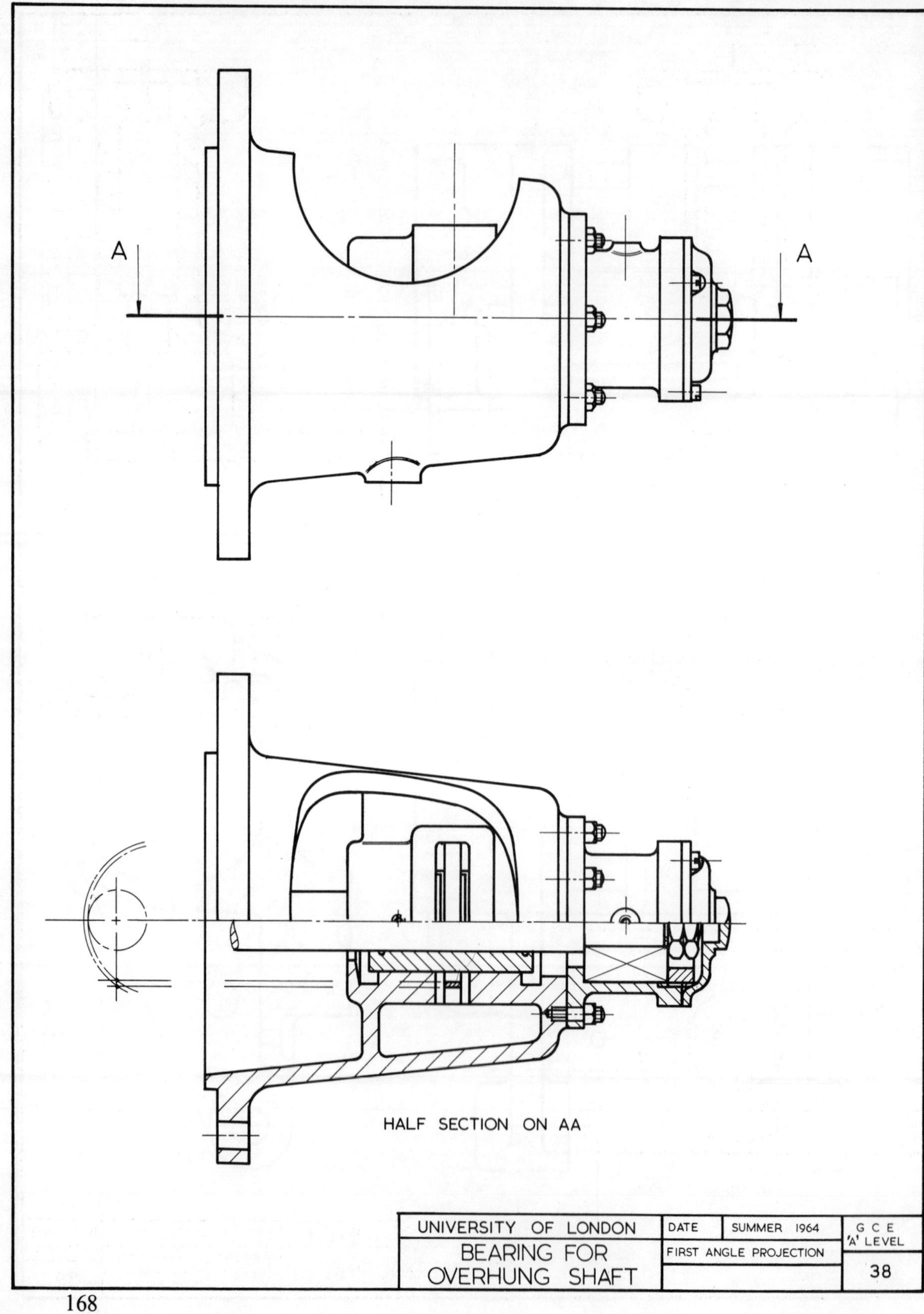
A
A
HALF SECTION ON AA
UNIVERSITY OF LONDON
BEARING FOR
OVERHUNG SHAFT
DATE
SUMMER 1964
FIRST ANGLE PROJECTION
G C E
'A' LEVEL
38

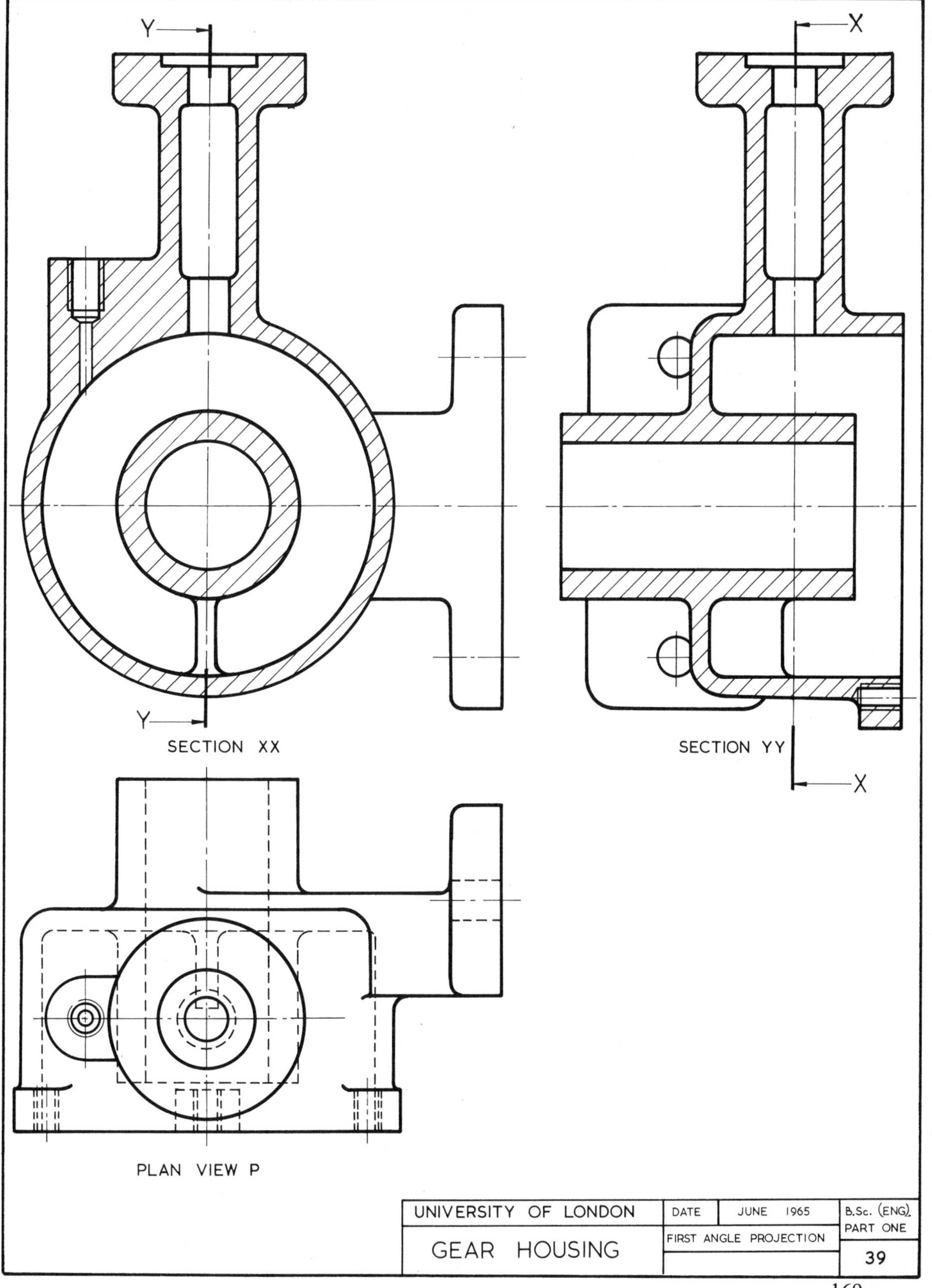
Y
X
Y
SECTION XX
SECTION YY
X
PLAN VIEW P
UNIVERSITY OF LONDON
DATE
JUNE 1965
B.Sc. (ENG).
PART ONE
GEAR HOUSING
FIRST ANGLE PROJECTION
39

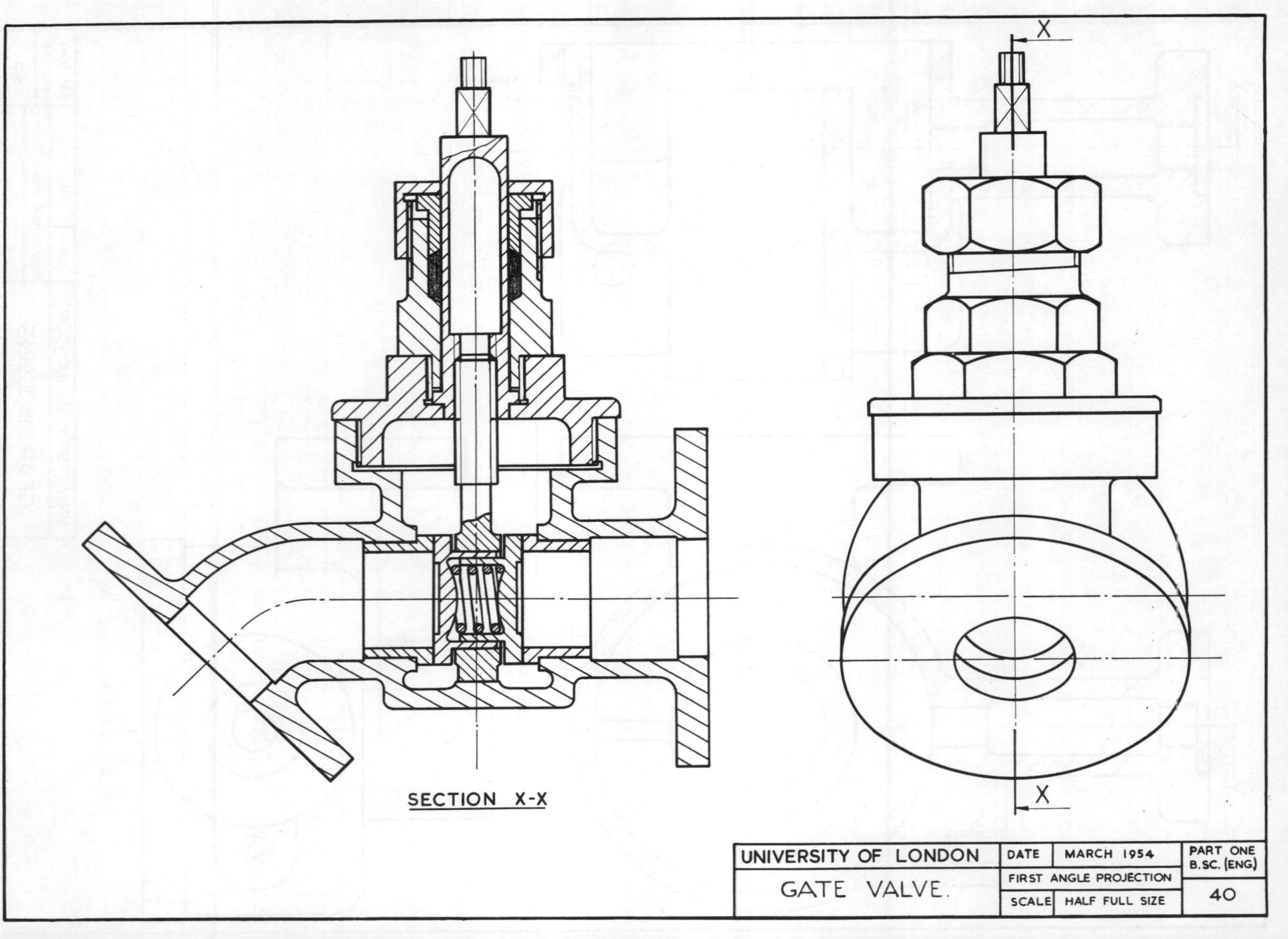
X
SECTION X-X
X
UNIVERSITY OF LONDON
GATE VALVE.
DATE
MARCH 1954
FIRST ANGLE PROJECTION
SCALE
HALF FULL SIZE
PART ONE
B.SC. (ENG)
40

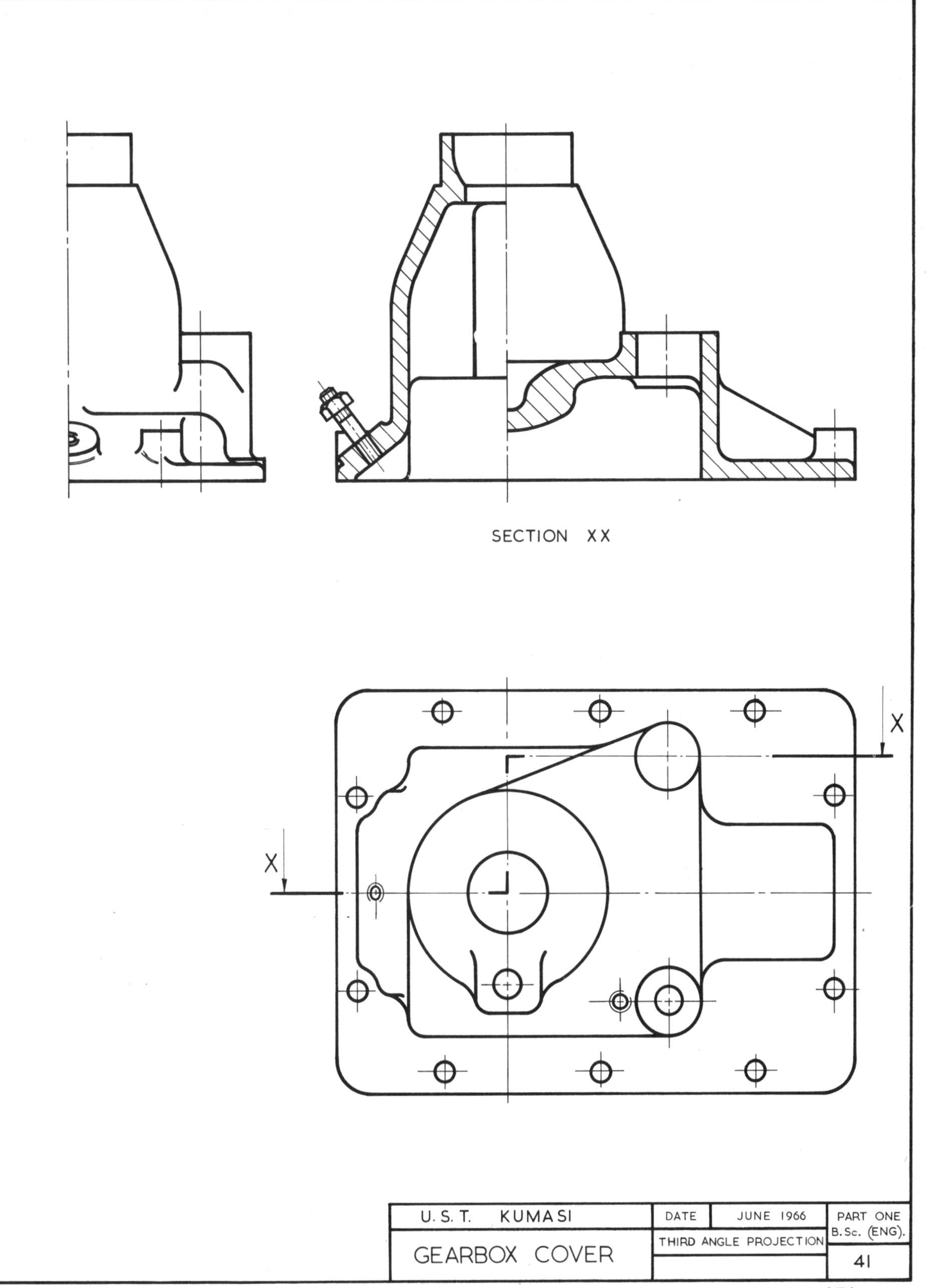
SECTION XX
X
X
U. S. T. KUMASI
DATE
JUNE 1966
PART ONE
B.Sc. (ENG).
GEARBOX COVER
THIRD ANGLE PROJECTION
41

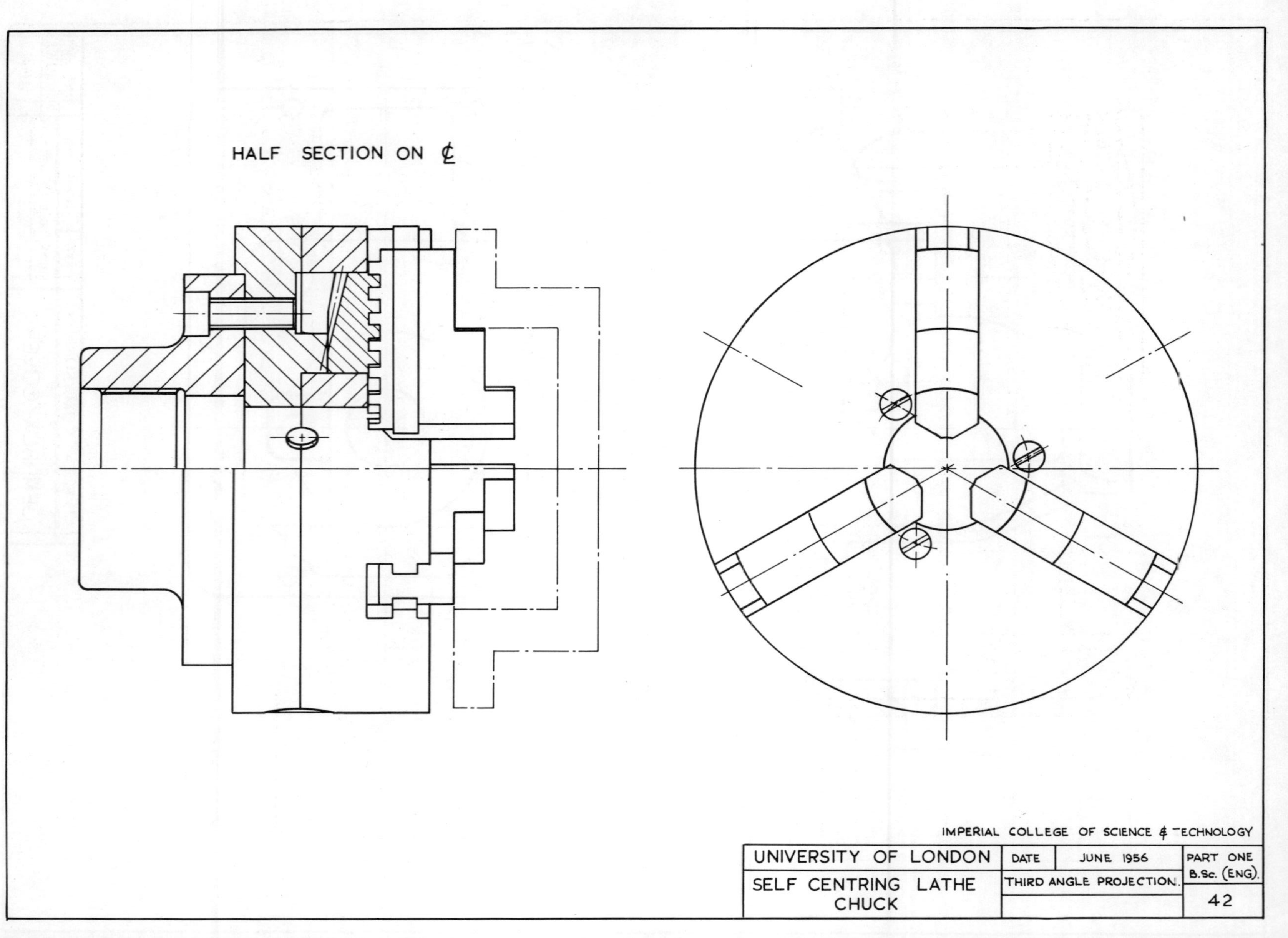
HALF SECTION ON ℄
IMPERIAL COLLEGE OF SCIENCE & TECHNOLOGY
UNIVERSITY OF LONDON
DATE
JUNE 1956
PART ONE
B.Sc. (ENG).
SELF CENTRING LATHE
CHUCK
THIRD ANGLE PROJECTION.
42

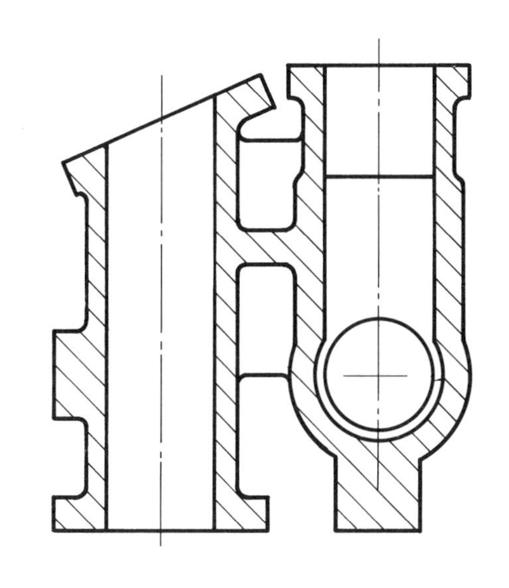

SECTION CD

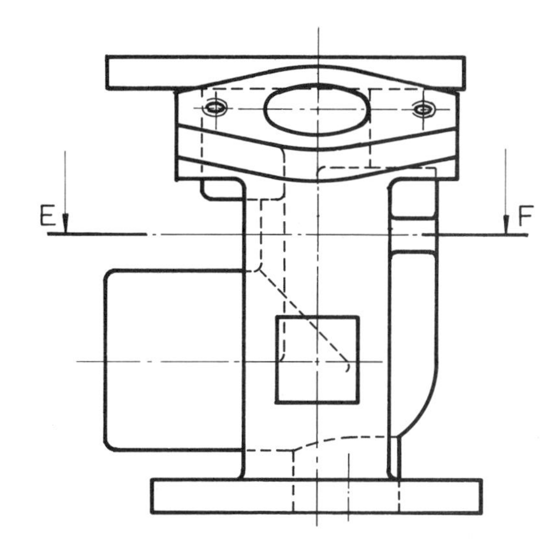

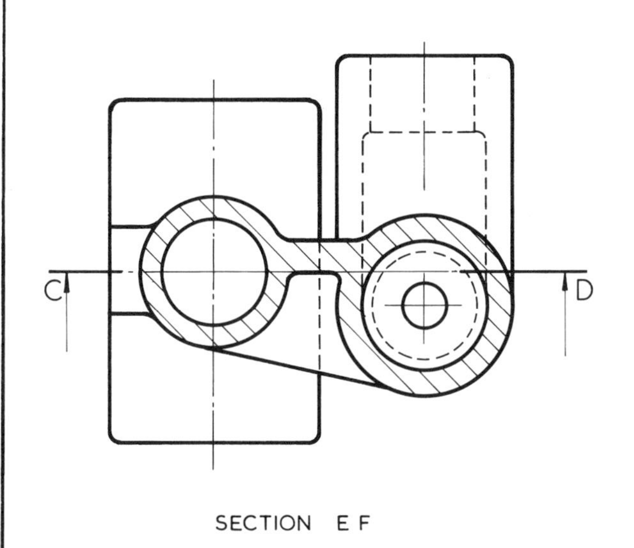

SECTION E F

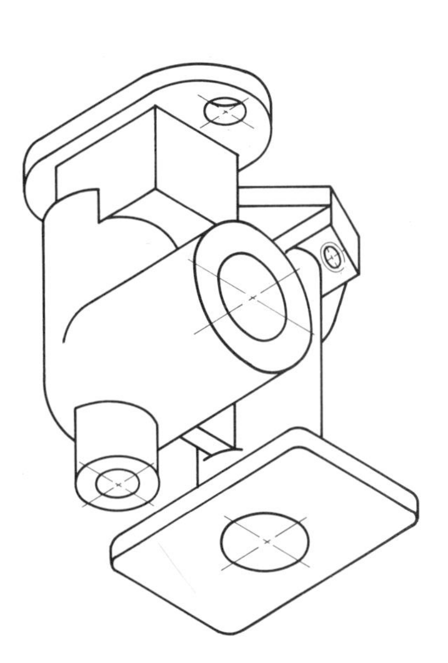

UNIVERSITY OF LONDON	DATE	JUNE 1963	PART ONE B.Sc.(ENG).
JUNCTION PIECE	FIRST ANGLE PROJECTION		43

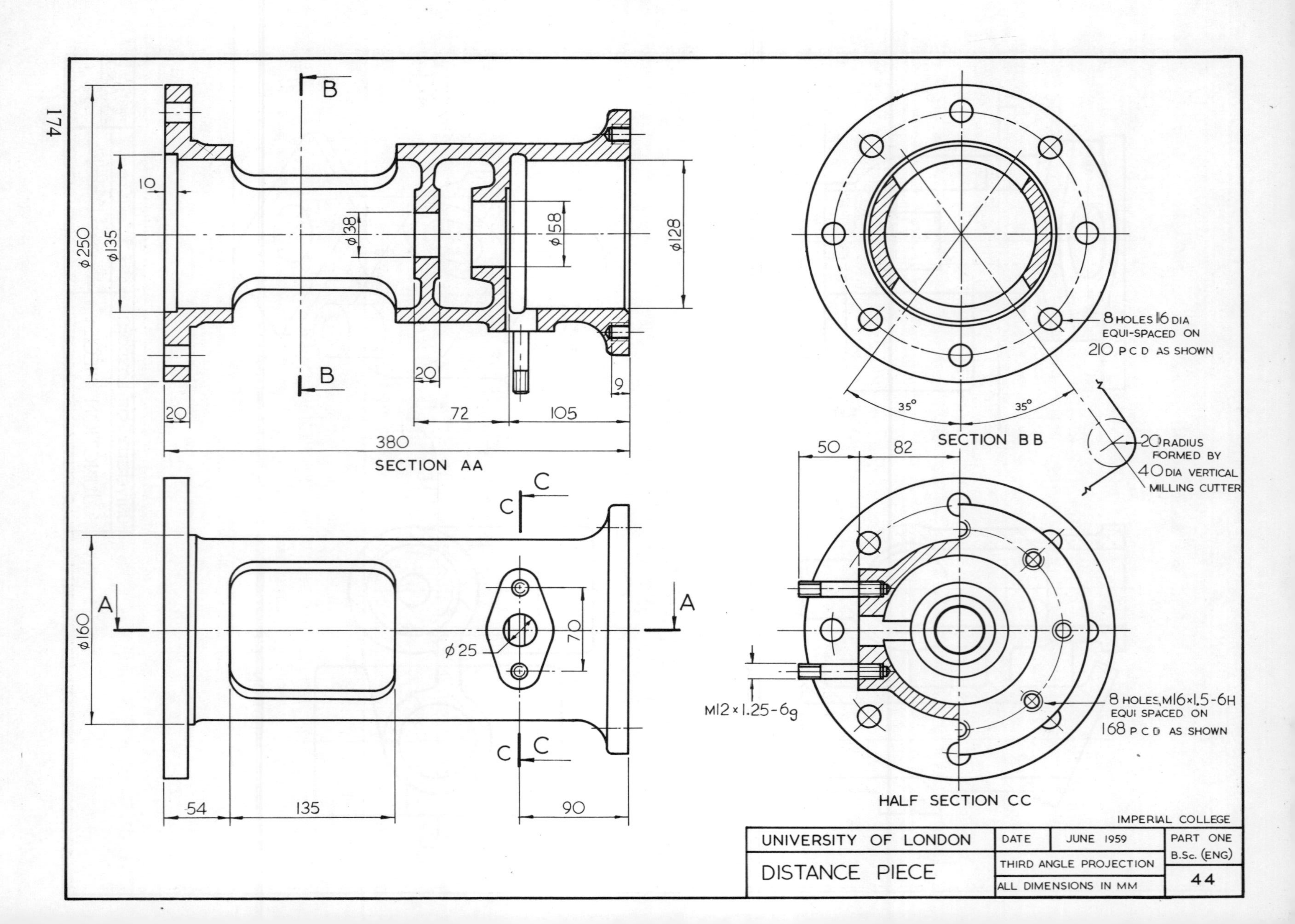
B
B
ϕ250
ϕ135
10
ϕ38
ϕ58
ϕ128
20
9
20
72
105
380
SECTION AA
8 HOLES 16 DIA
EQUI-SPACED ON
210 P C D AS SHOWN
35°
35°
SECTION B B
20 RADIUS
FORMED BY
40 DIA VERTICAL
MILLING CUTTER
50
82
C
C
C
C
A
A
ϕ160
ϕ25
70
54
135
90
M12 × 1.25-6g
8 HOLES, M16×1.5-6H
EQUI SPACED ON
168 P C D AS SHOWN
HALF SECTION CC
IMPERIAL COLLEGE
UNIVERSITY OF LONDON
DATE
JUNE 1959
PART ONE
B.Sc. (ENG)
DISTANCE PIECE
THIRD ANGLE PROJECTION
ALL DIMENSIONS IN MM
44

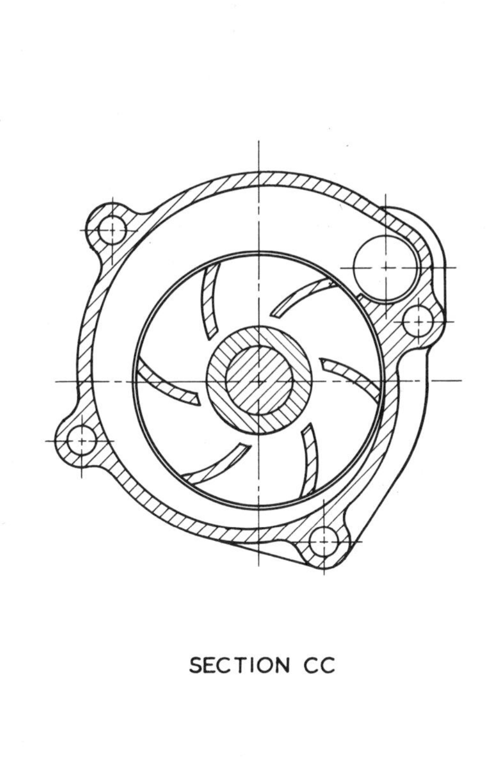

SECTION CC

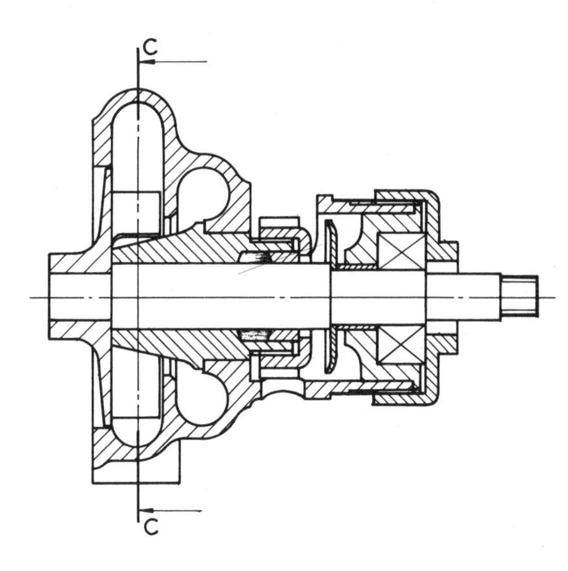

SECTION AA

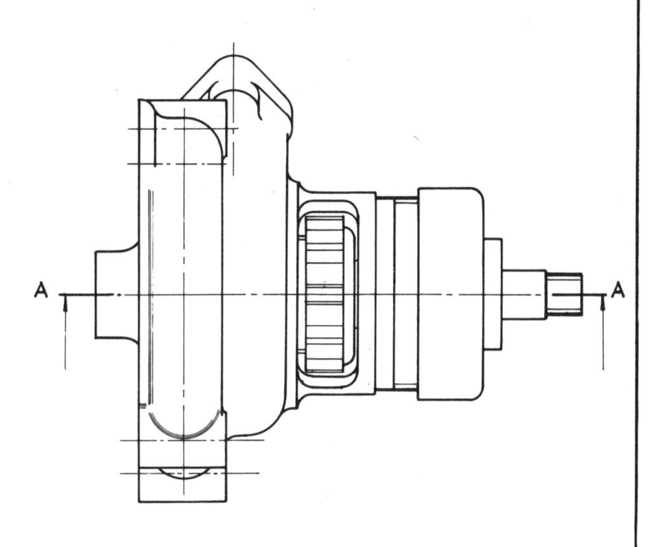

UNIVERSITY OF LONDON	DATE	JANUARY 1963	G.C.E A LEVEL
WATER PUMP	FIRST ANGLE PROJECTION		45

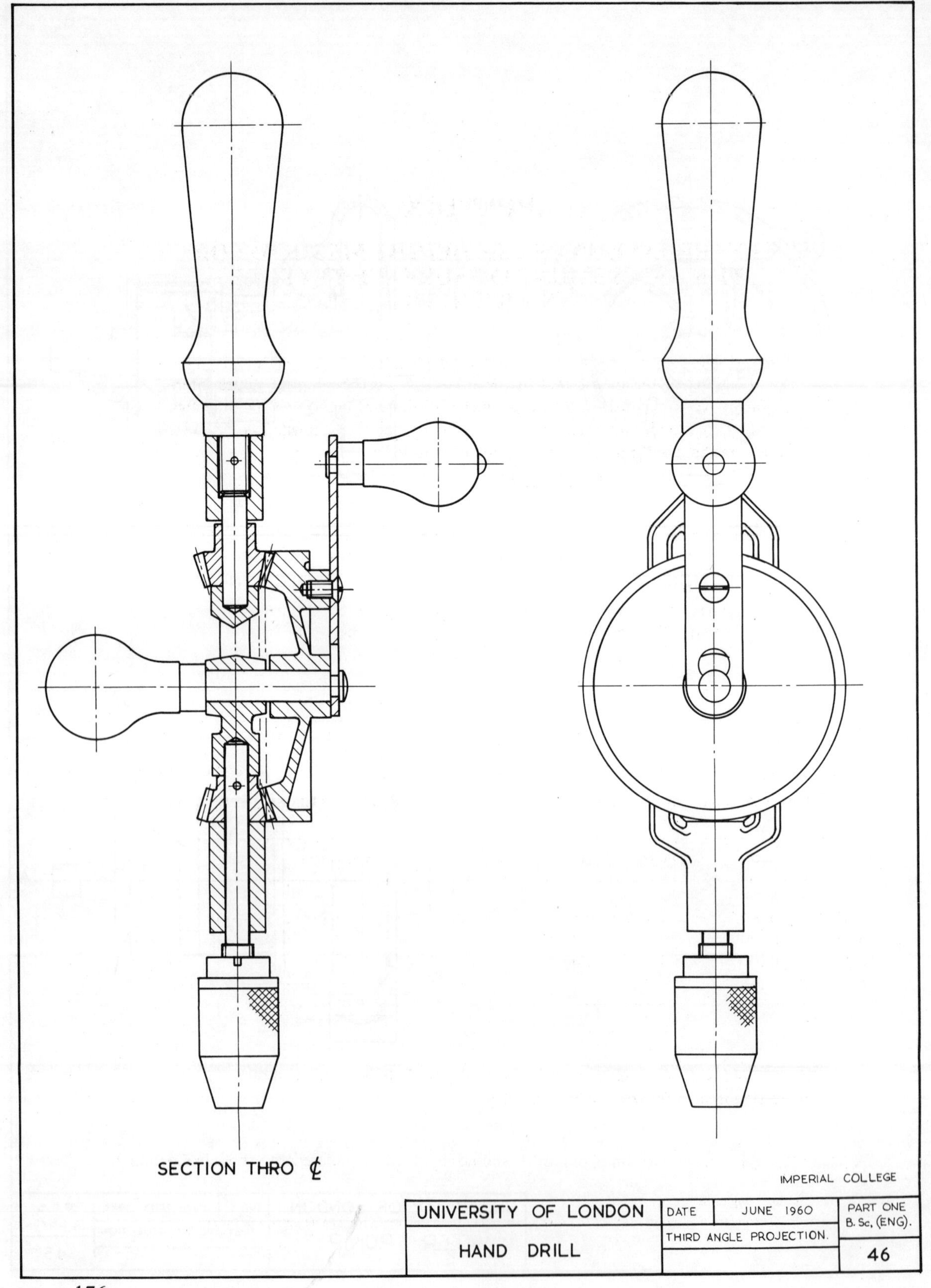
SECTION THRO ℄
IMPERIAL COLLEGE
UNIVERSITY OF LONDON
DATE
JUNE 1960
PART ONE
B. Sc, (ENG).
THIRD ANGLE PROJECTION.
HAND DRILL
46

APPENDIX

CENTRE-LINE-AVERAGE HEIGHT METHOD FOR THE ASSESSMENT OF SURFACE TEXTURE (See B S 1134)

The provisions of BS 1134 are equally valid using either inch or metric units. For designing in the metric system the following are the adopted values internationally recommended. For the convenience of those changing existing drawings and data to the metric system the related inch values are shown. The adopted values are those given in ISO Recommendation R 468*.

PREFERRED CLA VALUES
(Relative to Table 1 of BS 1134)

micrometre	microinch	micrometre	microinch	micrometre	microinch
0·025	1	0·4	16	6·3	250
0·050	2	0·8	32	12·5	500
0·1	4	1·6	63	25·0	1000
0·2	8	3·2	125		

STANDARD METER CUT-OFF VALUES
(Relative to Table 2 of BS 1134)

millimetre	inch	millimetre	inch
0·08	0·003	2·5	0·1
0·25	0·01	8·0	0·3
0·8	0·03	25·0	1·0

NOMINAL TIP RADIUS OF STYLUS
(Relative to Clause 7 of BS 1134)

micrometre	inch
2·5	0·0001
12·5	0·0005

In statements of surface texture the symbol μm shall be used to indicate expression in micrometres, i.e. 6·3 μm CLA.

NOTE. The diameters given in Column 1 should be used in preference to those in Column 2.

* ISO/R 468, 'Surface roughness'.

NOTE. The term 'micrometre' is used in place of the now obsolescent term 'micron'; 1 micrometre = 0·001 mm.

ISO METRIC PRECISION HEXAGON BOLTS AND SCREWS
(from BS 3692)

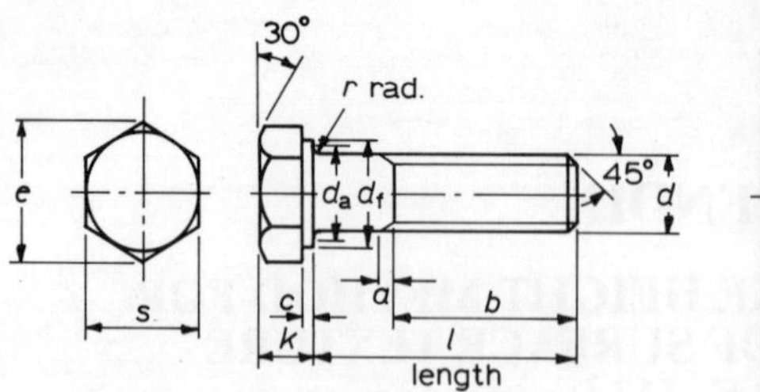

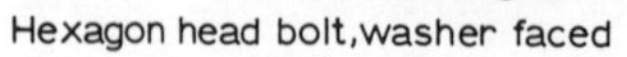
Hexagon head bolt, washer faced

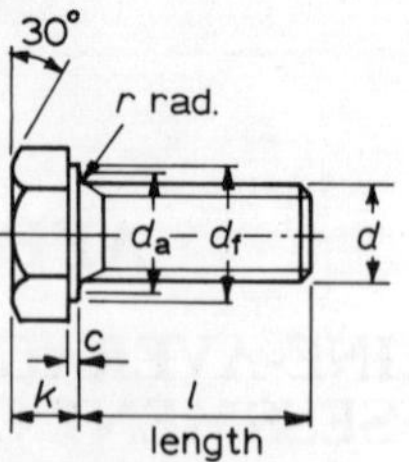

Hexagon head screw, washer faced

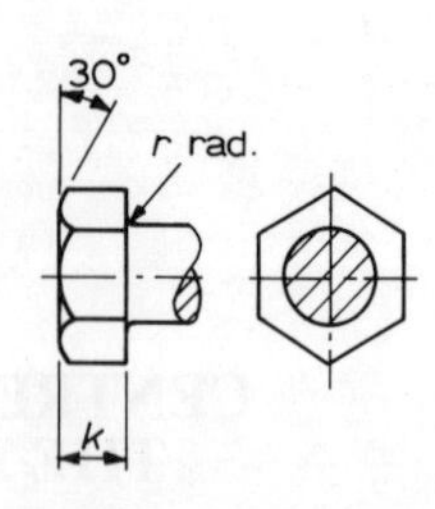

Full bearing head. Alternative type of head permissible on bolts and screws

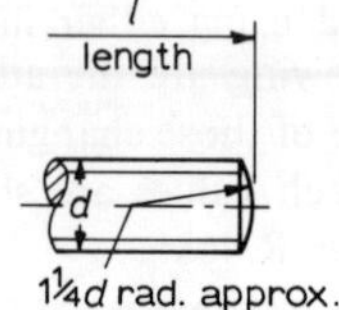

Rounded end

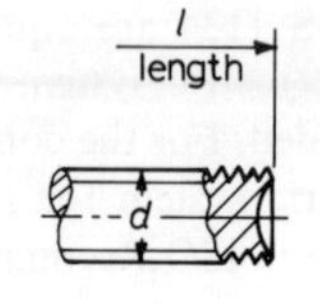

Rolled thread end

Alternative types of end permissible on bolts and screws

All dimensions in millimetres.

1	2	3	4	5	6	7	8	9	10	11	12	13	14	15	16	17
Nominal size and thread dia. *d*	Pitch of thread (coarse pitch series)	Thread runout *a*	Diameter of unthreaded shank *d*		Width across flats *s*		Width across corners *e*		Diameter of washer face *df*		Depth of washer face	Transition diameter* *da*	Radius under head* *r*		Height of head *k*	
		max.	max.	min.	max.	min.	max.	min.	max.	min.	*c*	max.	max.	min.	max.	min.
M3	0·5	1·2	3·0	2·86	5·5	5·38	6·4	6·08	5·08	4·83	0·1	3·6	0·3	0·1	2·125	1·875
M4	0·7	1·6	4·0	3·82	7·0	6·85	8·1	7·74	6·55	6·30	0·1	4·7	0·35	0·2	2·925	2·675
M5	0·8	2·0	5·0	4·82	8·0	7·85	9·2	8·87	7·55	7·30	0·2	5·7	0·35	0·2	3·650	3·35
M6	1	2·5	6·0	5·82	10·0	9·78	11·5	11·05	9·48	9·23	0·3	6·8	0·4	0·25	4·15	3·85
M8	1·25	3·0	8·0	7·78	13·0	12·73	15·0	14·38	12·43	12·18	0·4	9·2	0·6	0·4	5·65	5·35
M10	1·5	3·5	10·0	9·78	17·0	16·73	19·6	18·90	16·43	16·18	0·4	11·2	0·6	0·4	7·18	6·82
M12	1·75	4·0	12·0	11·73	19·0	18·67	21·9	22·10	18·37	18·12	0·4	14·2	1·1	0·6	8·18	7·82
M14	2	5·0	14·0	13·73	22·0	21·67	25·4	24·49	21·37	21·12	0·4	16·2	1·1	0·6	9·18	8·82
M16	2	5·0	16·0	15·73	24·0	23·67	27·7	26·75	23·27	23·02	0·4	18·2	1·1	0·6	10·18	9·82
M18	2·5	6·0	18·0	17·73	27·0	26·67	31·2	30·14	26·27	26·02	0·4	20·2	1·1	0·6	12·215	11·785
M20	2·5	6·0	20·0	19·67	30·0	29·67	34·6	33·53	29·27	28·80	0·4	22·4	1·2	0·8	13·215	12·785
M22	2·5	6·0	22·0	21·67	32·0	31·61	36·9	35·72	31·21	30·74	0·4	24·4	1·2	0·8	14·215	13·785
M24	3	7·0	24·0	23·67	36·0	35·38	41·6	39·98	34·98	34·51	0·5	26·4	1·2	0·8	15·215	14·785
M27	3	7·0	27·0	26·67	41·0	40·38	47·3	45·63	39·98	39·36	0·5	30·4	1·7	1·0	17·215	16·785
M30	3·5	8·0	30·0	29·67	46·0	45·38	53·1	51·28	44·98	44·36	0·5	33·4	1·7	1·0	19·26	18·74
M33	3·5	8·0	33·0	32·61	50·0	49·38	57·7	55·80	48·98	48·36	0·5	36·4	1·7	1·0	21·26	20·74
M36	4	10·0	36·0	35·61	55·0	54·26	63·5	61·31	53·86	53·24	0·5	39·4	1·7	1·0	23·26	22·74
M39	4	10·0	39·0	38·61	60·0	59·26	69·3	66·96	58·86	58·24	0·6	42·4	1·7	1·0	25·26	24·74
M42	4·5	11·0	42·0	41·61	65·0	64·26	75·1	72·61	63·76	63·04	0·6	45·6	1·8	1·2	26·26	25·74
M45	4·5	11·0	45·0	44·61	70·0	69·26	80·8	78·26	68·76	68·04	0·6	48·6	1·8	1·2	28·26	27·74
M48	5	12·0	48·0	47·61	75·0	74·26	86·6	83·91	73·76	73·04	0·6	52·6	2·3	1·6	30·26	29·74

NOTE. Sizes in bold type are preferred sizes.

* A true radius is not essential providing that the curve is smooth and lies wholly within the maximum radius, determined from the maximum transitional diameter, and the minimum radius specified.

ISO METRIC SCREW THREADS, GENERAL PLAN
(from BS 3643)

Dimensions in millimetres

NOTE. The basic major diameter is the same as the nominal diameter.

Nominal diameters			Pitches										
			Series with graded pitches		Series with constant pitches							
1st Choice	2nd Choice	3rd Choice	Coarse	Fine	3	2	1·5	1·25	1	0·75	0·5	0·35
1·6			0·35									
	1·8		0·35									
2			0·4									
	2·2		0·45									
1·5			0·45									0·35
3			0·5									0·35
	3·5		0·6									0·35
4			0·7								0·5	
	4·5		0·75								0·5	
5			0·8								0·5	
		5·5									0·5	
6			1							0·75		
		7	1							0·75		
8			1·25	1					1	0·75		
		9	1·25						1	0·75		
10			1·5	1·25				1·25	1	0·75		
		11	1·5						1	0·75		
12			1·75	1·25			1·5	1·25	1			
	14		2	1·5			1·5	1·25*	1			
		15					1·5		1			
16			2	1,			1·5		1			
		17					1·5		1			
	18		2·5	1·5		2	1·5		1			
20			2·5	1·5		2	1·5		1			
	22		2·5	1·5		2	1·5		1			
24			3	2		2	1·5		1			
		25				2	1·5		1			
		26					1·5					
	27		3	2		2	1·5		1			
		28				2	1·5		1			
30			3·5	2	(3)	2	1·5		1			
		32				2	1·5					
	33		3·5	2	(3)	2	1·5					
		35.					1·5					
36			4	3	3	2	1·5					
		38					1·5					
	39		4	3	3	2	1·5					
1st Choice	2nd Choice	3rd Choice	Coarse	Fine	3	2	1·5	1·25	1	0·75	0·5	0·35
			Series with graded pitches		Series with constant pitches							
Nominal diameter			Pitches									

Avoid as far as possible pitches in brackets.
*Pitch 1·25 of diameter 14: to be used only for spark plugs for engines.
Diameter 35: to be used only for locking nuts for bearings.

NOTE. The pitches enclosed in the bold frame, together with the corresponding 1st and 2nd choice nominal diameters in Columns 1 and 2 are those combinations which have been established by ISO recommendations as a selected 'coarse' and 'fine' series for screws, bolts and nuts and other common threaded fasteners. For these applications Tolerance Class 6H/6g has been adopted.

ISO METRIC SCREW THREADS, GENERAL PLAN

Dimensions in millimetres

Nominal diameters			Pitches									
			Series with graded pitches		Series with constant pitches							
1st Choice	2nd Choice	3rd Choice	Coarse	Fine	6	4	3	2	1·5			
		40					3	2	1·5			
42			4·5	3		4	3	2	1·5			
	45		4·6	3		4	3	2	1·5			
48			5	3		4	3	2	1·5			
		50	5	3		4	3	2	1·5			
	52		5	3		4	3	2	1·5			
		55				4	3	2	1·5			
56			5·5	4		4	3	2	1·5			
		58				4	3	2	1·5			
	60		5·5	4		4	3	2	1·5			
		62				4	3	2	1·5			
64			6	4		4	3	2	1·5			
		65				4	3	2	1·5			
	68		6	4		4	3	2	1·5			
		70			6	4	3	2	1·5			
72					6	4	3	2	1·5			
		75				4	3	2	1·5			
	76				6	4	3	2	1·5			
		78						2				
80					6	4	3	2	1·5			
		82						2				
	85				6	4	3	2				
90					6	4	3	2				
		95			6	4	3	2				
100					6	4	3	2				
	105				6	4	3	2				
110					6	4	3	2				
	115				6	4	3	2				
	120				6	4	3	2				
125					6	4	3	2				
	130				6	4	3	2				
		135			6	4	3	2				
140					6	4	3	2				

Nominal diameter			Series with graded pitches		Series with constant pitches							
					Pitches							
1st Choice	2nd Choice	3rd Choice	Coarse	Fine	6	4	3	2	1·5			
		145			6	4	3	2				
	150				6	4	3	2				
		155			6	4	3					
160					6	4	3					
		165			6	4	3					
	170				6	4	3					
		175			6	4	3					
180					6	4	3					
		185			6	4	3					
	190				6	4	3					
		195			6	4	3					
200					6	4	3					
		205			6	4	3					
	210				6	4	3					
		215			6	4	3					
220					6	4	3					
		225			6	4	3					
		230			6	4	3					
		235			6	4	3					
	240				6	4	3					
		245			6	4	3					
250					6	4	3					
		255			6	4						
	260				6	4						
		265			6	4						
		270			6	4						
		275			6	4						
280					6	4						
		285			6	4						
		290			6	4						
		295			6	4						
	300				6	4						

NOTE 1. Where a coarse series thread is required in sizes above 68 mm, the 6 mm pitch from the constant pitch series should be used.

NOTE 2. Where a fine series thread is required in sizes above 68 mm, the 4 mm pitch from the constant pitch series should be used.

NOTE 3. Where a fine series thread is required in sizes below 8 mm, the constant pitch series should be used.

ISO METRIC PRECISION HEXAGON NUTS AND THIN NUTS
(from BS 3692)

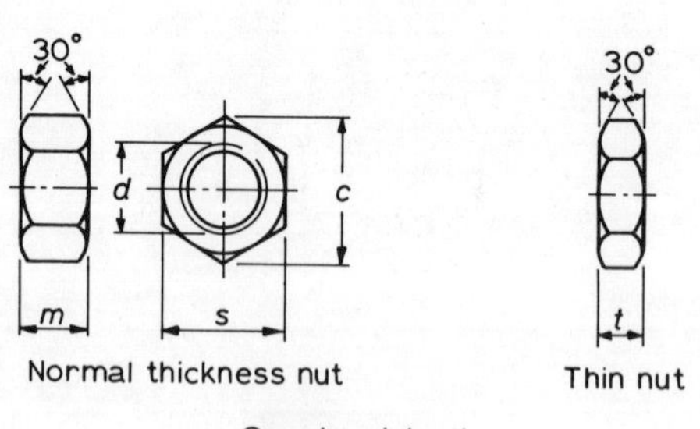

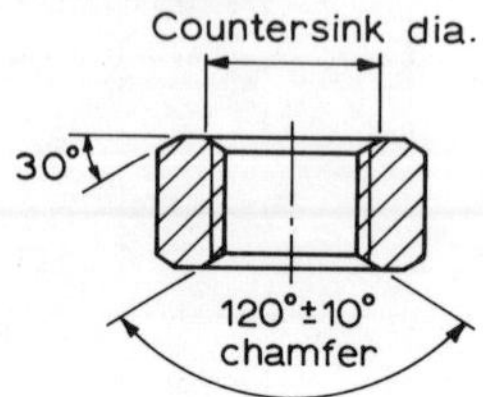

Enlarged view of nut countersink

All dimensions in millimetres.

1	2	3	4	5	6	7	8	9	10	11	12
Nominal size and thread diameter *d*	**Pitch of thread (coarse pitch series)**	**Width across flats** *s*		**Width across corners** *e*		**Thickness of normal nut** *m*		**Tolerance on squareness of thread to face of nut***	**Eccentricity of hexagon**	**Thickness of thin nut** *t*	
		max.	**min.**	**max.**	**min.**	**max.**	**min.**	**max.**	**max.**	**max.**	**min.**
M3	0·5	5·50	5·38	6·40	6·08	2·40	2·15	0·09	0·14	—	—
M4	0·7	7·00	6·85	8·10	7·74	3·20	2·90	0·11	0·18	—	—
M5	0·8	8·00	7·85	9·20	8·87	4·00	3·70	0·13	0·18	—	—
M6	1	10·00	9·78	11·50	11·05	5·00	4·70	0·17	0·18	—	—
M8	1·25	13·00	12·73	15·00	14·38	6·50	6·14	0·22	0·22	5·0	4·70
M10	1·5	17·00	16·73	19·60	18·90	8·00	7·64	0·29	0·22	6·0	5·70
M12	1·75	19·00	18·67	21·90	21·10	10·00	9·64	0·32	0·27	7·0	6·64
M14	2	22·00	21·67	25·40	24·49	11·00	10·57	0·37	0·27	8·0	7·64
M16	2	24·00	23·67	27·70	26·75	13·00	12·57	0·41	0·27	8·0	7·64
M18	2·5	27·00	26·67	31·20	30·14	15·00	14·57	0·46	0·27	9·0	8·64
M20	2·5	30·00	29·67	34·60	33·53	16·00	15·57	0·51	0·33	9·0	8·64
M22	2·5	32·00	31·61	36·90	35·72	18·00	17·57	0·54	0·33	10·0	9·64
M24	3	36·00	35·38	41·60	39·98	19·00	18·48	0·61	0·33	10·0	9·64
M27	3	41·00	40·38	47·30	45·63	22·00	21·48	0·70	0·33	12·0	11·57
M30	3·5	46·00	45·38	53·10	51·28	24·00	23·48	0·78	0·33	12·0	11·57
M33	3·5	50·00	49·38	57·70	55·80	26·00	25·48	0·85	0·39	14·0	13·57
M36	4	55·00	54·26	63·50	61·31	29·00	28·48	0·94	0·39	14·0	13·57
M39	4	60·00	59·26	69·30	66·96	31·00	30·38	1·03	0·39	16·0	15·57
M42	4·5	65·00	64·26	75·10	72·61	34·00	33·38	1·11	0·39	16·0	15·57
M45	4·5	70·00	69·26	80·80	78·26	36·00	35·38	1·20	0·39	18·0	17·57
M48	5	75·00	74·26	86·60	83·91	38·00	37·38	1·29	0·39	18·0	17·57

NOTE. Sizes in bold type are preferred sizes.

* See Appendix A.

CLEARANCE HOLES FOR METRIC BOLTS
(from BS 4186)

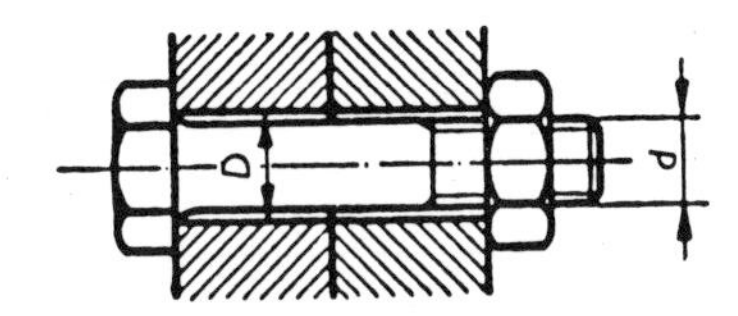

Dimensions in millimetres

Thread diameter *d*	Clearance holes *D*		
	Series		
	fine	medium	coarse
1·6	1·7	1·8	2
2	2·2	2·4	2·6
2·5	2·7	2·9	3·1
3	3·2	3·4	3·6
4	4·3	4·5	4·8
5	5·3	5·5	5·8
6	6·4	6·6	7
7	7·4	7·6	8
8	8·4	9	10
10	10·5	11	12
12	13	14	15
14	15	26	17
16	17	18	19
18	19	20	21
20	21	22	24
22	23	24	26
24	25	26	28
27	28	30	32
30	31	33	35
33	34	36	38
36	37	39	42
39	40	42	45

Tolerances. If it is considered necessary to apply a tolerance to the above hole diameters, the following should be adopted:

fine series	H 12
medium series	H 13
coarse series	H 14

WASHERS FOR HEXAGON BOLTS AND NUTS
METRIC SERIES
(from BS 4320)

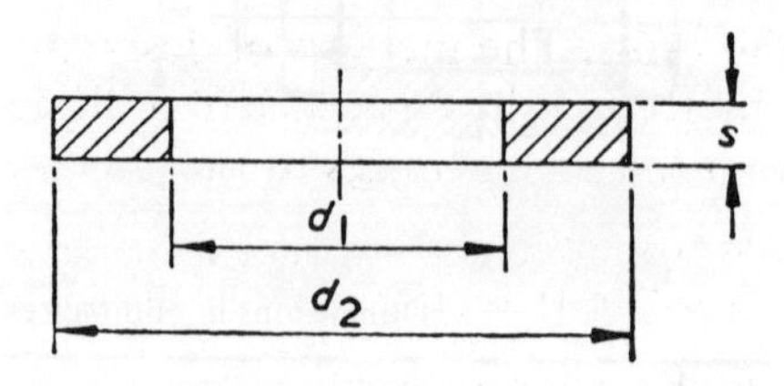

Clearance hole d_1		Diameter of washer d_2 for width across flats			Thickness of washer s	For thread diameters
for bright washers	for black washers	normal	small	large		
mm	mm	mm	mm	mm	mm	mm
1·7	—	4	—	—	0·3	1·6
2·2	—	5	—	—	0·3	2
2·7	—	6·5	—	—	0·5	2·5
3·2	—	7	—	—	0·5	3
4·3	—	9	—	—	0·8	4
5·3	5·5	10	—	—	1	5
6·4	6·6	12·5	—	—	1·6	6
7·4	7·6	14	—	—	1·6	7
8·4	9	17	15·5	21	1·6*	8
10·5	11	21	18	24	2	10
13	14	24	21	28	2·5	12
15	16	28	24	30	2·5	14
17	18	30	28	34	3	16
19	20	34	30	37	3	18
21	22	37	34	39	3	20
23	24	39	37	44	3	22
25	26	44	39	50	4	24
28	30	50	44	56	4	27
31	33	56	50	60	4	30
34	36	60	56	66	5	33
37	39	66	60	72	5	36
40	42	72	66	77	6	39

* reads 1·6.

BASIC SIZES FOR METAL, SHEET, STRIP AND WIRE
(from BS 4391)

Range of sizes. The range of sizes is from 0·020 to 25 mm.
Preference for sizes. In selecting sizes, preference should be given to sizes in the R10, R20 or R40 series in that order.
Method of designating the thickness. The method of designating the thickness of sheet or diameter of wire is to be by stating the basic size in millimetres, followed, if desired, by the letter U to indicate that this size belongs to the ISO metric series.

NOTE. The equivalent inch values for sizes above 0·25 mm are given to an accuracy close to, or better than one part in one thousand. This accuracy would be appropriate to practical limits of size associated with a tolerance of ± 1 per cent of the size. For sizes smaller than 0·25 mm, five places of decimals appear adequate for any likely method of direct measurement in inches. The true millimetre basic size should be used, if it desired to compute limits of size in any alternative characteristic, such as mass or electrical resistance.

TABLE OF BASIC SIZES

NOTE. Preference should be given to sizes in the R10, R20, R40 series, in that order.

Basic sizes			Basic sizes			Basic sizes		
millimetres			millimetres			millimetres		
R10	R20	R40	R10	R20	R40	R10	R20	R40
0·020	0·020	0·020	0·063	0·063	0·063	0·200	0·200	0·200
		0·021			0·067			0·212
	0·022	0·022		0·071	0·071		0·224	0·224
		0·024			0·075			0·236
0·025	0·025	0·025	0·080	0·080	0·080	0·250	0·250	0·250
		0·026			0·085			0·265
	0·028	0·028		0·090	0·090		0·280	0·280
		0·030			0·096			0·300
0·032	0·032	0·032	0·100	0·100	0·100	0·315	0·315	0·315
		0·034			0·106			0·335
	0·036	0·036		0·112	0·112		0·355	0·355
		0·038			0·118			0·375
0·040	0·040	0·040	0·125	0·125	0·125	0·400	0·400	0·400
		0·042			0·132			0·425
	0·045	0·045		0·140	0·140		0·450	0·450
		0·048			0·150			0·475
0·050	0·050	0·050	0·160	0·160	0·160	0·500	0·500	0·500
		0·053			0·170			0·530
	0·056	0·056		0·180	0·180		0·560	0·560
		0·060			0·190			0·600
0·063	0·063	0·063	0·200	0·200	0·200	0·630	0·630	0·630

CONVENTIONAL REPRESENTATION OF COMMON FEATURES
(from BS 308)

SUBJECT | CONVENTION

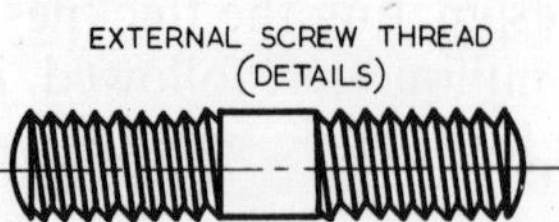

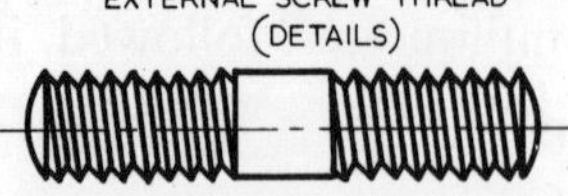
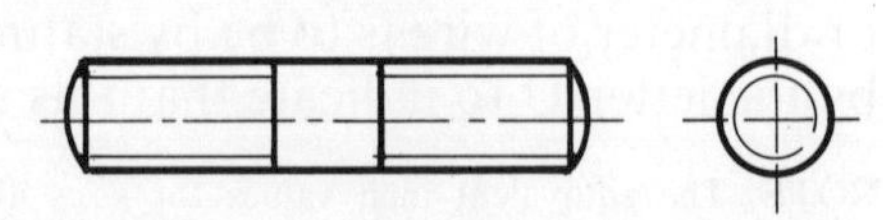

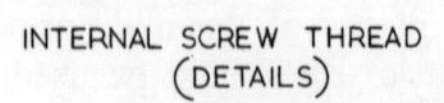

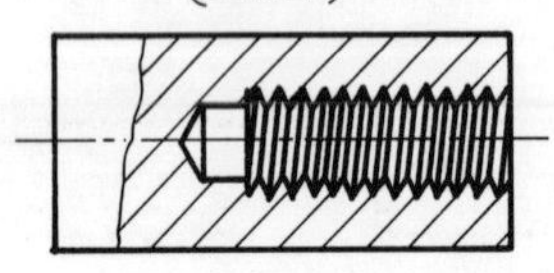
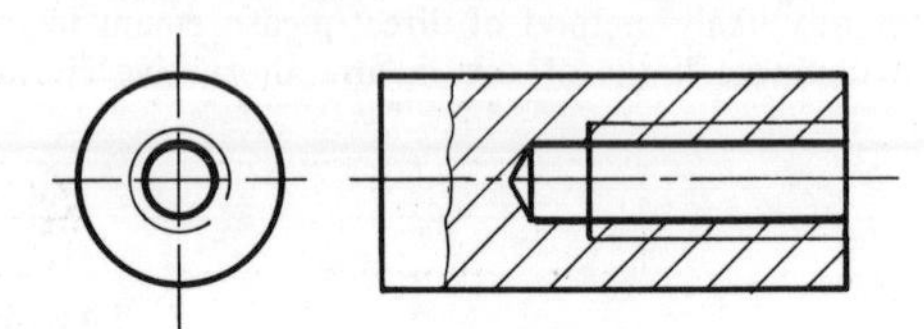

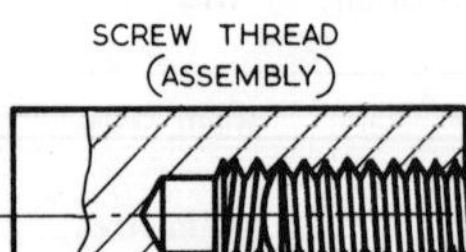

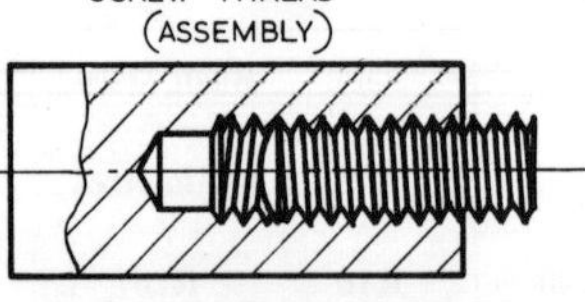
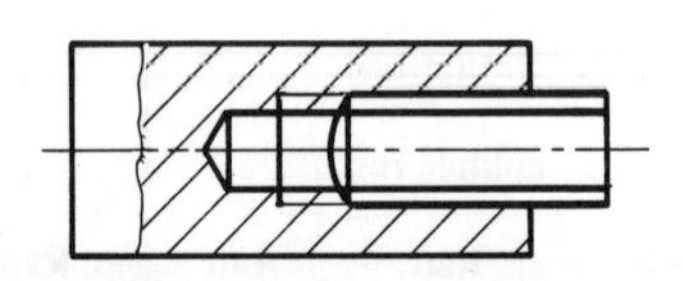

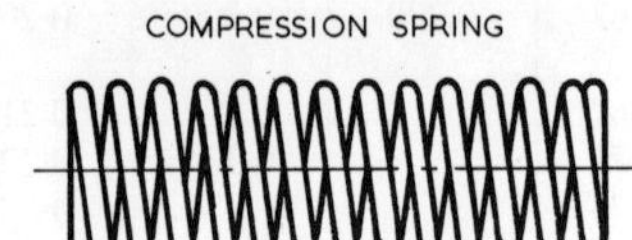

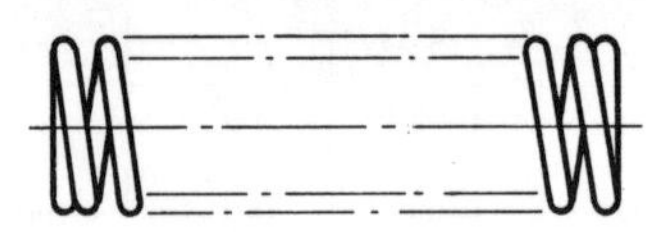

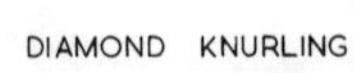

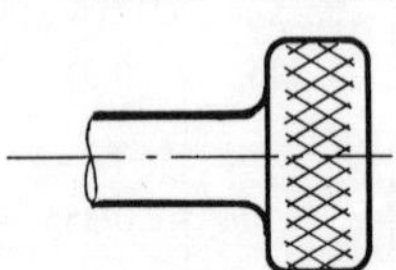
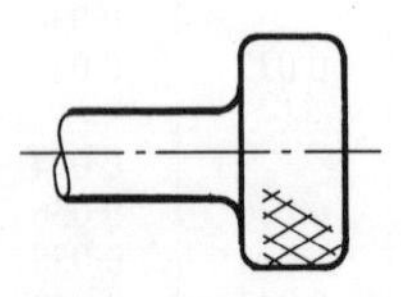

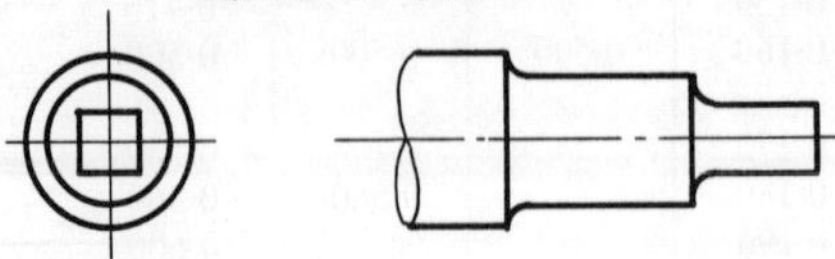

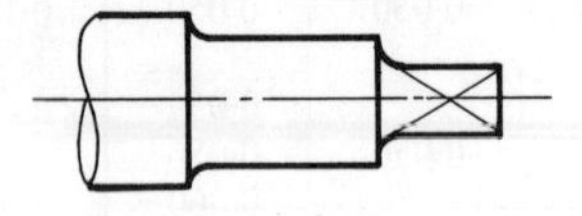

BEARINGS

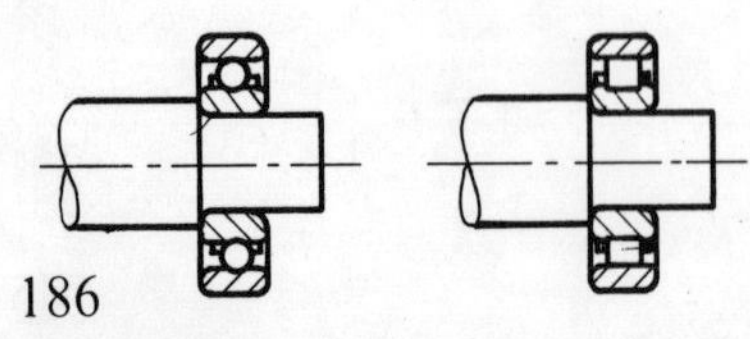
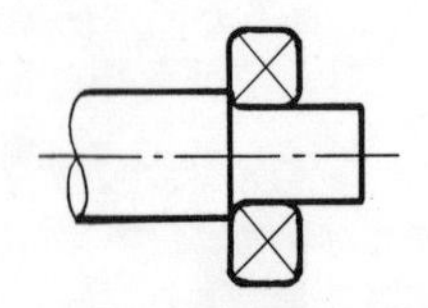

SUBJECT | CONVENTION

BREAK LINES
ROUND (SOLID)

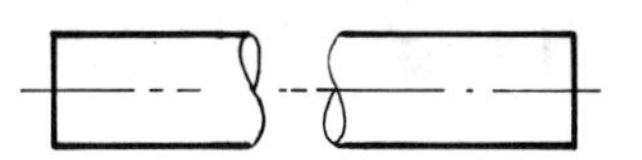

BREAK LINES
ROUND (TUBULAR)

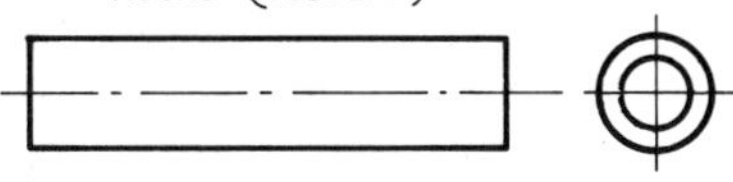

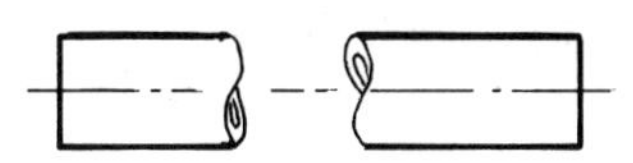

SPUR GEAR
(DETAILS)

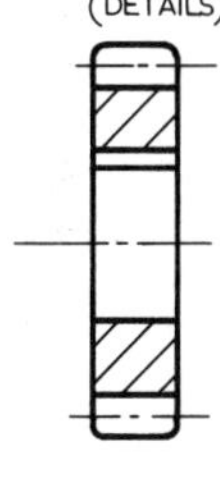

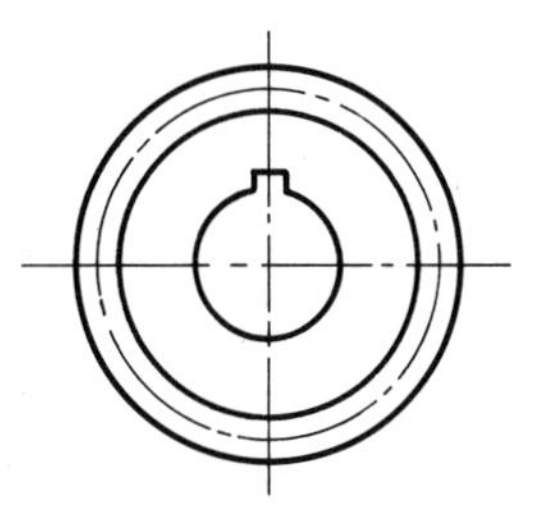

TABLE OF ABBREVIATIONS FOR USE ON DRAWINGS
(from BS 308)

The following abbreviations should be used on drawings when required. Abbreviations are the same in the singular and plural. Although capital letters are shown, lower-case letters may be used where appropriate. Full stops are not used except when the abbreviation makes a word, e.g. the abbreviation for the word 'figure'.

Further recognized abbreviations are listed in other British Standards.

GENERAL ENGINEERING TERMS

Term	*Abbreviation*
Across flats	A/F
Assembly	ASSY
Centres	CRS
Centre line	L
Chamfered	CHAM
Cheese head	CH HD
Countersunk	CSK
Countersunk head	CSK HD
Counterbore	C'BORE
Cylinder or cylindrical	CYL
Diameter (in a note)	DIA
Diameter (preceding a dimension)	Ø
Drawing	DRG
Figure	FIG.
Hexagon	HEX
Hexagon head	HEX HD
Hydraulic	HYD
Insulated or insulation	INSUL
Left hand	LH
Long	LG
Material	MATL
Maximum	MAX
Minimum	MIN
Number	NO.
Pattern number	PATT NO.
Pitch circle diameter	PCD
Pneumatic	PNEU
Radius (preceding a dimension)	R*
Required	REQD
Right hand	RH
Round head	RD HD
Screwed	SCR
Sheet	SH
Sketch	SK
Specification	SPEC
Spherical diameter (preceding a dimension)	SPHERE Ø
Spherical radius (preceding a dimension)	SPHERE R
Spotface	S'FACE
Square (in a note)	SQ
Square (preceding a dimension)	□
Standard	STD
Threads per inch	TPI
Undercut	U'CUT
Volume	VOL
Weight	WT

*CAPITAL LETTER ONLY

TERMS RELATING TO DIMENSIONS AND TOLERANCES

Term	*Abbreviation*
Datum Datum system Datum dimensions	DATUM
True position, or true profile, dimension in conjunction with positional, or profile tolerances	TP
Angularity tolerance	ANG TOL
Concentricity tolerance	CONC TOL
Cylindricity tolerance	CYL TOL
Flatness tolerance	FLAT TOL
Parallelism tolerance	PAT TOL
Positional tolerance	POSN TOL
Roundness tolerance	RD TOL
Straightness tolerance	STR TOL
Squareness tolerance	SQ TOL
Symmetry tolerance	SYM TOL
Tolerance zone (profiles)	TOL ZONE
Maximum material condition	MMC
Full indicated movement	FIM